ENERGY SECURITY AND GEOPOLITICS IN THE ARCTIC

Challenges and Opportunities in the 21st Century

ENERGY SECURITY AND GEOPOLITICS IN THE ARCTIC

Challenges and Opportunities in the 21st Century

Editor

Hooman Peimani

Energy Studies Institute, National University of Singapore

ENERGY STUDIES INSTITUTE
National University of Singapore

World Scientific

NEW JERSEY • LONDON • SINGAPORE • BEIJING • SHANGHAI • HONG KONG • TAIPEI • CHENNAI

Published by

World Scientific Publishing Co. Pte. Ltd.
5 Toh Tuck Link, Singapore 596224
USA office: 27 Warren Street, Suite 401-402, Hackensack, NJ 07601
UK office: 57 Shelton Street, Covent Garden, London WC2H 9HE

British Library Cataloguing-in-Publication Data
A catalogue record for this book is available from the British Library.

The image "Northwest Passage–Northern Sea Route," seen on the front cover, is taken from the link: http://maps.grida.no/go/graphic/arctic-sea-routes-northern-sea-route-and-northwest-passage; Hugo Alhenius, UNEP/GRID-Arendal.

ENERGY SECURITY AND GEOPOLITICS IN THE ARCTIC
Challenges and Opportunities in the 21st Century

ISBN-13 978-981-4401-46-3 (pbk)
ISBN-10 981-4401-46-3 (pbk)

In-house Editor: Ms Sandhya Venkatesh

Typeset by Stallion Press
Email: enquiries@stallionpress.com

Printed by FuIsland Offset Printing (S) Pte Ltd Singapore

DISCLAIMER

The views and ideas expressed by the contributing authors of this book do not necessarily represent those of the Energy Studies Institute (ESI) and the publication of their chapters does not imply that ESI subscribes to or agrees with their expressed views and ideas.

CONTENTS

LIST OF ACRONYMS

AC	Arctic Council
AARI	Arctic and Antarctic Research Institute
ACIA	Arctic Climate Impact Assessment
ALDE	Alliance of Liberals and Democrats for Europe
AMAP	Arctic Monitoring and Assessment Program
AO	Arctic Ocean
APEC	Asia-Pacific Economic Cooperation
ASEAN	Association of Southeast Asian Nations
ATF	Arctic Task Force
BAU	Business As Usual
BB	Billion barrels
bcm	Billion cubic meters
BEAC	Barents Euro-Arctic Council
Bq	Becquerel, unit of measuring radionuclide content
BRC	Barents Regional Council
CAFE	Conservational Arctic Flora and Fauna
CBSS	Council of the Baltic Sea States
CCS	Carbon Capture and Sequestration
DDT	Dichloro Diphenyl Trichloroethane
EC	European Commission
ECCS	Emergency Core Cooling System
EEZ	Exclusive Economic Zone
ENSO	El Niño-Southern Oscillation
EP	European Parliament
EU	European Union
FPU	Floating Production Unit

G8	The Group of Eight, a forum for the governments of eight major economies
GHGs	Greenhouse gases
Gt	Gigatons
HM	Heavy metals
IASC	International Arctic Science Committee
IEA	International Energy Agency
IMO	International Maritime Organization
IPCC	Intergovernmental Panel on Climate Change
IOR	Indian Ocean Region
ISA	UN International Seabed Authority
LNG	Liquefied natural gas
mb/d	Million barrels per day
METI	Ministry of Economics, Trade and Industry
MPC	Maximum Permissible Concentrations
MPEC	Maximum Permissible Ecological Concentrations
Msl	Mean sea level
mtoe	Million tons of oil equivalent
NATO	North Atlantic Treaty Organization
NCM	Nordic Council of Ministers
NEAFC	Northeast Atlantic Fisheries Commission
NGOs	Non-Governmental Organizations
NSR	Northern Sea Route
NSW	Northern Sea Way
NWP	North-West Passage
PDA	Previously Disputed Area
PCP	Phencyclidine
POP	Persistent organic pollutants
PS	Polluting Substances
PSSAs	Particularly Sensitive Sea Areas
SAO	Search and Rescue in the Arctic
SCM	Standard cubic metres
SDAG	Shtokman Development AG
SEA	Sea Economic Activities
SLR	Sea level rise

SOC	Steady organic compounds
STF	Arctic Special Task Force
SWIPA	Snow, Water, Ice and Permafrost in the Arctic
tcf	trillion cubic feet
tcm	trillion cubic meters
UK	United Kingdom
UN	United Nations
UNCLCS	United Nations Commission on the Limits of the Continental Shelf
UNCLOS	United Nations Convention on the Law of the Sea
UNSC	United Nations Security Council
USA	United States of America
USGC	US Geological Survey
USSR	Union of Soviet Socialist Republics
°C	Celsius degrees
°F	Fahrenheit degrees

LIST OF TABLES AND FIGURES

Tables

Figures

ACKNOWLEDGMENTS

I acknowledge with appreciation the contribution of the Energy Security Division at the Energy Studies Institute of the National University of Singapore, which helped organize a workshop on the Arctic in 2011 that laid the ground for preparing this book as well as other related activities throughout that year and in early 2012. I thank its researchers, namely, Mr. Nur Azha Putra, Mr. Nicholas Koh Chung Wei, Ms. Mallika Naguran and Ms. Su Liying. In particular, Ms. Su's assistance with various aspects of the formatting and organizing of the book was critical for its completion. I would also like to express appreciation for the administrative support provided by ESI's Assistant Manager, Ms. Jan Lui.

ABOUT THE CONTRIBUTORS

1. Dr. Hooman Peimani

Dr. Hooman Peimani is a Principal Fellow and Head of the Energy Security Division at the Energy Studies Institute of the National University of Singapore. He specializes in energy (energy security) and security (regional/international and human security), particularly relating to South and West Asia, the Middle East, the Asia-Pacific region and the Arctic.

Having worked at different senior capacities, Dr. Peimani's diversified employment engagements in North America, Europe and West Asia have included working for academic entities (e.g., Geneva School of Diplomacy & International Relations) and non-academic ones, such as energy corporations (e.g., Nord Stream, Switzerland), private and public organizations (e.g., Canadian government) and national and international institutions (e.g., UNICEF, UNCHR, UNRISD and WHO). His employment experience includes over 20 years of research activities working for research institutes such as Jane's Defence/UK, Geneva Centre for the Democratic Control of Armed Forces, Centre for International Cooperation and Security/UK and Arab Petroleum Research Center/France. He has extensive field experience in West and South Asia and the Asia/Pacific region.

Added to his years of experience as a freelance journalist working for many newspapers (e.g., *South China Morning Post* and *Moscow Times*) and news agencies (e.g., *Inter Press Service*/Berlin, *International Relations and Security Network/*Zurich and *Eurasianet/* New York), he has contributed as an expert to many newspapers (e.g., *The Wall Street Journal*) and news agencies (e.g., *Asahi Shimbon*, *Bloomberg*, *Kyodo News*, *Platts*, *Reuters* and *UPI*) and/or been quoted by them. Since the 1990s, he has made regular contributions to many radio/TV programs (e.g., *Deutsche Welle*, *BBC*, *CNA* and *CNBC*). His publications include 10 books, 27 chapters in books, over 200 journal/newspaper articles, scores of government/UN documents/reports and several book reviews.

2. Dr. Shawn J. Marshall

Dr. Shawn Marshall is a Professor in the Department of Geography at the University of Calgary, Canada. He is also a Canada Research Chair in Climate Change, and a Research Associate with the Canadian Institute for Advanced Research (CIfAR) Earth System Evolution Program.

Dr. Marshall studies glacier and ice sheet dynamics and global glacier response to climate change, including field studies in mountain and polar regions. This involves three-dimensional numerical modelling of ice sheet dynamics in Greenland, Iceland and the Canadian Rocky Mountains. Dr. Marshall is the current Acting Executive Director of the Arctic Institute of North America. He is a contributing author to the IPCC and a lead author of the recent Arctic Monitoring and Assessment Report (AMAP). He has published widely on glacier response to climate change, sea level rise, and snow and ice in the climate system, including the recent textbook *The Cryosphere.*

Dr. Marshall studied Engineering Physics at the University of Toronto and went on to do his PhD in Geophysics at the University of British Columbia.

3. Professor Viacheslav Mikhailovich Makeev

Professor Viacheslav Mikhailovich Makeev is a professor in the Geography Department of the State Polar Academy in Russia. He majors in Arctic research, Geography, Geology, Paleogeography, Ecology and Nature Management.

He has been Chair of the Nature Management Department at the State Polar Academy since 2000. He also heads the ecological security section of the Arctic Expertise Council across the Arctic regions of the Council of the Russian Federation (Parliament).

Professor Mikhailovich graduated from the Geography Department of St. Petersburg University in 1963. He holds a PhD in geomorphology and paleogeography. He has participated in 25 polar expeditions to the Arctic and spent a winter season in the Antarctic.

4. Associate Professor Nadezhda Klimovna Kharlampyeva

Associate Professor Nadezhda Kharlampyeva teaches courses on Contemporary Problems of World Politics, Arctic Political Methodology, Transnational Corporations and World Political Issues, International Organizations and Political Systems Processes at St Petersburg University in Russia. She also teaches in the Regional Economics Department of the Nevsky Language and Cultural Institute.

Over the years, she has worked for other educational entities, including the educational section of Sakha-Russian Diamond Company, St Petersburg's Secretariat of the Northern Forum NGO and the City Investment Agency of the St Petersburg Government.

Associate Professor Kharlampyeva is the editor of the Russian Natural Science Academy's *Arctic Idea Journal*. She currently directs research projects on Arctic policy sponsored by Russian grants. Professor Kharlampyeva has participated and organized numerous national and international conferences and seminars. She is an author of the monograph, *North-West Federal Area: External Affairs* (2008) (co-authored with N. Markushina) and more than 40 science articles on Arctic cooperation and Arctic policy.

With a PhD in International Relations and Foreign Policy, she is a member of the Russian International Relations Studies Group (RISA MGIMO) and the International Arctic Social Science Association (IASSA).

5. Dr. Arild Moe

Dr. Arild Moe is a Deputy Director/Senior Research Fellow with the Fridtjof Nansen Institute at Lysaker, near Oslo, Norway. Most of his research has been devoted to Russia and the region, with specific interest in the energy sector, mainly the oil and gas industry. Dr. Moe has also conducted research on offshore activities in the Barents Sea, Russian climate politics, and also Russian–Norwegian relations.

In 1992, he was a consultant for the European Bank for Reconstruction and Development's Energy Sector Technical Assistance Project for the Baltic countries. Presently, Dr. Moe is directing a long-term research project on energy developments in Russia and the Caspian Region with participants from several Norwegian and foreign institutes. He is also engaged in studies on broader Arctic policy issues as well as Norwegian policy in the high north. Dr. Moe has written extensively, including peer-reviewed articles and book chapters. His most recent research was published in *Polar Geography, Eurasian Geography and Economics*, and the *Arctic Review on Law and Politics.*

Dr. Moe has undergraduate degrees in Political Science, Russian Language and Public Law and a PhD in Political Science from the University of Oslo.

6. Dr. Danila Bochkarev

Dr. Danila Bochkarev is with the Global Security team of the Brussels-based East-West Institute. Before joining this Institute, he was an Inbev scholar of EU-Russia relations at the Louvain-la Neuve and Leuven Universities in Belgium. He also worked on China and Central Asian affairs at the European Parliament.

Dr. Bochkarev studied international relations, history and political economy in Paris, London and Nizhniy Novgorod (Russia). In addition to his PhD, he has several post-graduate qualifications in international relations.

He is the author of many academic publications and policy papers, and regularly contributes to international media and policy debates on the EU, Russian and Eurasian energy, security and foreign policy issues. Dr. Bochkarev has several academic and professional distinctions, including being a Royal Dutch Shell and French Government scholar. He is a native Russian speaker and is fluent in English and French.

7. Commander Neil Gadihoke

Commander Neil Gadihoke is a Research Fellow at India's premier think tank, the National Maritime Foundation (NMF), New Delhi. He has over 21 years of service in the Indian Navy and has held a wide array of operational, training and staff billets. He has an MSc in Telecommunications and an MSc in Defence and Security Studies. He also holds an MBA in Human Resource Management.

Commander Gadihoke has travelled to more than 25 countries and is a keen follower of international geopolitics. He has published articles on geopolitics and security in various journals. His present research areas include the geopolitics of the IOR (Indian Ocean Region), climate change and its implications for security, and the unfolding future of the Arctic. He has conducted research covering issues such as "Climate Change: Implications for the Indian Navy" and "The Emerging Scenario of Alternative Energies in the IOR".

He is a graduate of the US Navy Staff College, Rhode Island, USA where he was awarded the Zimmerman Gray Writing Award for the best research paper written by an international student during the 2004–2005 academic year. Subsequently, Commander Gadihoke was awarded the Commendation Card by the Chief of the Indian Navy while serving at the Navy's Directorate of Strategy, Concepts and Transformation, where he was a member of the core team that drafted the Indian Navy's Maritime Strategy. He also served as Naval Assistant to the Commander in Chief of the Strategic Forces Command and as Joint Director of Naval Telecommunications.

8. Professor Hiroshi Ohta

Professor Hiroshi Ohta is a Professor at the School of International Liberal Studies, Waseda University in Tokyo, where he teaches Political Science, Environmental Studies, Japanese Foreign Policy and Global Governance. His research themes include global environmental politics and policies, the concept of global governance, Japanese climate change policy, comparative studies of regional fisheries management regimes, and civil society and political theory.

Professor Ohta is a member of the Executive Board of the Japanese Chapter of the United Nations Studies Association. Previously, he sat on the Japanese Association of International Relations' Board of Advisors.

His recent relevant publications include: "Japanese Climate Change Policy: Moving beyond the Kyoto Process" in Hans G. Brauch *et al.* (eds), *Coping with Global Environmental Change, Disasters and Security: Threats, Challenges, Vulnerabilities and Risks* (Berlin: Springer, 2010: 1381–1391) and "A Small Leap Forward: Regional

Cooperation for Tackling the Problems of the Environment and Natural Resources in Northeast Asia'', in Martina Timmermann and Jituso Tsuchiyama (eds.), *Institutionalizing Northeast Asia: Regional Steps towards Global Governance* (New York: United Nations University, 2008: 297–315).

Professor Ohta received his PhD in International Relations from the Department of Political Science at the Graduate School of Arts and Sciences of Columbia University in New York City.

ABOUT THE ENERGY STUDIES INSTITUTE

Established as an autonomous research institute within the National University of Singapore, the Energy Studies Institute (ESI) is a policy-oriented research institute that seeks to produce independent and high quality research that stimulates and informs debate on energy issues of key interest to policy makers, industries and the community. ESI endeavours to advance the understanding of local, regional and global energy issues through independent research and analyses aimed at addressing, informing and influencing public opinion and policies. Research activities at ESI address issues that are at the intersection of energy economics, energy and the environment, and energy security.

1

INTRODUCTION

Hooman Peimani

Global warming is surely the most formidable challenge of the 21st century as its consequences have affected and will continue to affect the survival of our planet. Although there are still some scientists, representing a minority view, who deny this as a global phenomenon, major climatic changes reflected, for example, in frequent droughts, forest fires, floods and rising sea levels in different parts of the world, including the Asia-Pacific region, serve as evidence to the contrary. Among those who acknowledge it as a global reality, some attribute it to natural and periodic changes, which have been in place for millions of years creating consecutive long or short phases of global warming and global cooling. Hence, the ongoing global warming is simply yet another episode of warming based on the natural pattern of climatic change being a characteristic of our planet caused mainly by terrestrial and celestial factors. Unnatural factors caused by the activities of the human race are not the causes of this phenomenon although they may have some local implications, as the view holds. Yet, global warming has now been accepted by the majority of scientists concerned with climatic changes and just about all governments as a real and expanding phenomenon with short-and long-term dire consequences should it continue. Accordingly, the human-made emission of greenhouse gases (GHG) is the main cause of this phenomenon. Within this context, the main problem is CO_2, which is generated

Fig. 1.1. Arctic Region.

Source: Author's creation based on Arctic blank map by Radostaw Botev. http://en.wikipedia.org/wiki/File:Arctic blank map.png

primarily by the heavy global consumption of fossil fuels (oil, gas, and coal), a phenomenon that has been in place over the last two centuries. It has had various manifestations in different parts of the world, including destruction of forests caused by frequent fires, rapid depletion of water resources, desertification and rapid/permanent melting ice. The latter has been manifested in the rapid melting of glaciers in different parts of the world (e.g., those of the Himalaya) and the melting of the Arctic ice.

1.1. CONTEXT FOR THE ARCTIC'S DEBATE: CHALLENGES

Hence, global warming underlies the Arctic debate, which is the focus of this book, and has various implications ranging from environmental to energy security. While melting of some land-based large concentrations of ice (glaciers) may have, at least initially, local implications, the melting of huge floating Arctic ice is a major development with global implications. In other words, this phenomenon, if it continues, will pose formidable challenges, given its disastrous short-and long-term environmental consequences, and will seriously question the survival of this planet and its living creatures in worst-case scenarios. Apart from worsening global warming, which will substantially diminish the world's largest cooling source, the melting of the Arctic ice will eventually trigger a process affecting all glaciers to varying extents, especially the one in Greenland, which is the world's largest. This process will prompt sea levels to rise over time at a rate faster than what we are witnessing today. The International Panel on Climate Change (IPCC) projections are alarming enough for the benchmark year of 2100. Thus far, the rise is projected to be 14.61 inches (37.11 centimetres) and 19.13 inches (48.59 centimetres) for A1B[1] and A1FI[2] scenarios.[3] The IPCC estimated that the Arctic could be free of summer sea ice somewhere between 2050 and 2080,[4] and 10% (possibly as much as 50%) of the Arctic tundra could be replaced by forests by 2100.[5] Apocalyptical results could be expected in the case of a total melting of the Arctic ice, which, in turn, would melt the Greenland ice and raise the sea level by 7 meters.[6]

1.2. OPPORTUNITIES: EXPLORATION OF RESOURCES AND NAVIGATION

The gradual melting of the Arctic is surely serious because of all its negative environmental implications, not only for the Arctic region, but also for the rest of the world. The resulting damage to the environment will eventually and inevitably affect other sectors, including the economic, political, and social ones. Yet, in the short-and possibly medium-term, the gradual melting of the ice could provide certain opportunities, such as energy extraction and mining, as the Arctic region is rich in fossil fuels, particularly oil and gas, as well as minerals. Directly related to this, the possibility of ice-free sea routes, for at least a few months of the year, through the Arctic could stimulate intercontinental trade. In short, this unfortunate environmental event could have certain benefits for a limited period of time, as will be briefly discussed in the following sections.

1.2.1. *Energy*

Based on the currently available data, the Arctic is the last known region of the world with vast untapped reserves of oil and gas as well as large coal deposits. Of course, there are various estimates of the size of the resources under the ice and in the deep sea, which can be confirmed only when major exploration projects begin. Moreover, it is still uncertain, though there are estimates, as to just how much of these resources are commercially retrievable given the high cost of deep sea oil/gas extraction in such a climatically-hostile environment. Nevertheless, based on mean estimates, the entire region — consisting of those areas within the Arctic states' territorial boundaries — and the rest of the region, including the disputed areas, has at least 90 billion barrels of undiscovered oil.[7] The mean estimates for undiscovered natural gas and natural gas liquids are 47.2891 trillion cubic meters and 44 billion barrels, respectively.[8]

1.2.2. *Minerals*

The Arctic contains large quantities of various minerals, including precious metals and stones. They include nickel, copper, tin, uranium, phosphate, lead, zinc, gold, apatite, alumina, iron ore, mica, titanium, phlogopite, vermiculite, niobium, tantalite, quartz, fluorite, barite, calcite, amethyst, amazonite, and diamonds.[9] Their estimated value of US$1.5–2 trillion clearly indicates the reason for their importance to the regional states.[10] In this regard, the following examples are noteworthy as they reflect the regional and global significance of the Arctic minerals. In the American Arctic part, Alaska has major reserves of minerals, plus large reserves of zinc and lead. The Red Dog mine, for instance, is a zinc/lead mine located within the boundaries of the Red Dog Mine census-designated place. The world's largest producer of zinc in its heyday, and also the largest mineral producer in the state of Alaska in 2009, accounting for 53.3% of its annual production, it now has the world's largest zinc reserves. At the end of 2009, the most recent year for which data is available, the mine had reserves of 67.66 million tons of zinc of 16.6–20% grade and 67.66 million tons of lead of 4.4–5.4% grade.[11]

Canada's Arctic section is also rich in minerals. For example, its iron deposits are significant for their size and production capabilities, as evident in the case of an undertaking by the Baffinland Iron Mines Corporation. Currently, this corporation is developing a large open pit iron mine in the Mary River area of Baffin Island, Nunavut, Canada, which consists of nine high-grade iron ore deposits.[12] Scheduled to make its first commercial ore shipments in September 2014, the mine is expected to produce 18 million tons of iron ore per year for 20 years.[13]

The European Arctic section is rich in minerals similar to the North American section. Iron ore is also in abundance here. As a Greater Arctic country, Sweden contains the region's largest iron-bearing area in Swedish Lapland, of which the total discovered reserves (mainly magnetite and hematite) are estimated to be 2,413 million metric tons.[14] Norway also has large deposits of iron ore, including about 1,000 million tons in its Northern part (Sør-Varanger) and a large reserve of iron ore has also recently been surveyed at Kolari in northwest Finland.[15]

1.2.3. *Northern sea route and North-West passage*

The melting of the Arctic ice, if it continues, will result in part of the Arctic becoming ice-free for a few months each year in the future. If this happens, it will be possible to navigate between the Atlantic and Pacific Oceans via the Arctic and thus through the Northern Sea Route via the Russian Arctic and the Northwest Passage via the Canadian Arctic, which will be significantly shorter than the currently-used Southern Sea Route through the Suez Canal and the

Strait of Malacca.Today, the travelling distance for a ship from Murmansk in Russia to Yokohama in Japan via the Suez Canal is 20,554 kilometres, but it would be only 9,232 kilometres through the Arctic (Northern Sea Route).[16] Likewise, the distance between South Korea and the Netherlands is about 20,320 kilometres via the Southern Sea Route, while it would be only approximately 14,720 kilometres via the Northern Sea Route.[17]

Such a possibility could, and likely would, facilitate, indeed, encourage trade between Europe, North America and the Northeastern part of Asia (Asia-Pacific region) by reducing the cost of inter-continental exports/imports. Heavy sea traffic through the Arctic for a few months of the year would certainly contribute to large-scale air and water pollution, further contributing to the region's warming with obvious negative environmental impacts on the Arctic and subsequently the rest of the world. Apart from its environmental dimension, such traffic could potentially and significantly affect sea traffic through the Southern Sea Route as Asian, European, and North American countries with access to the Arctic sea routes would likely have a strong incentive to conduct part of their inter-continental trade through them during its few months of availability.

Today, no one can predict the real extent of cargo and possibly oil and liquefied natural gas (LNG) tanker traffic, through these routes to determine the impact on the Southern Route. To start with, large-scale commercial use of the Northern Sea Route, which is comparatively more developed than the North-West Passage and thus more likely to be used at a relatively significant scale in the foreseeable future, requires the route remaining ice-free for a few months a year or at least a situation close to being ice-free requiring limited use of icebreakers to keep the route navigable. This would enable unhindered navigation of large cargo ships and/or sea tankers. Contingent on the continuation of global warming, such scenarios are predicted to happen sometime in the future, but the approximate date is disputed by various scholars. Thus, estimates for an ice-free Northern Sea Route in summer, prolonging the sailing season from the current 20–30 days to about 120 days range from as early as 2013 to as late as 2080, but most estimates fall between 2040 and 2060.[18] According to research carried out by Cambridge University, as presented by Professor Peter Wadhams in 2011, the sea route could well be available even sooner than 2013 as the whole North Pole could be "ice-free in just a few years' time, all gone by the summer of 2015."[19] The total disappearance of the Arctic Ocean by 2015 is of course a nightmare scenario given the calamitous consequences for which no country is prepared.

One should hope that the required global efforts, hinging on a substantial decrease of GHG emissions, will take place soon enough to end or at least slow down global warming to avoid its catastrophic effects, including a significant and steady melting of the Arctic ice. Nevertheless, if the melting of the Arctic ice continues, the availability of the Northern Sea Route for a few months a year will certainly decrease cargo/tanker traffic via the Southern Sea Route for those countries with access to the Northern Sea Route, with obvious negative economic implications for the countries along the Southern Sea Route. There should also be a decrease in the geostrategic importance of this route and, by default, its respective countries. However, the mentioned negative consequences should be limited given that the bulk of intercontinental trade will have to be conducted through the Southern Sea Route in any case, given that the use of the Arctic routes, including the Northern Sea Route, would not be a geographically feasible option for such trade involving Southern European, Southern American, West and South Asian and many Asia-Pacific countries. In particular, the large and growing intraregional trade, such as that from the Asia-Pacific region, will still have to be done through the Southern Sea Route regardless of the availability of the Arctic sea routes.

1.2.4. *Immediate and aspiring beneficiaries*

Assuming that the current environmentally disastrous trend continues, the melting of the Arctic ice will open doors for certain economic activities of not just regional, but also global significance, including oil and gas extraction, mining and intercontinental navigation. Without a doubt, the main immediate beneficiaries of such activities are the five littoral states, namely: Canada, Denmark/Greenland,

Norway, Russia and the United States. By virtue of sharing parts of the region, they have direct access to it as well as opportunities to expand into the unchartered and/or disputed parts of the Arctic. Moreover, the bulk of the region's known mineral and energy (oil and gas) resources fall within their territories and exclusive economic zones (EEZ). These resources are unevenly distributed and leave certain countries, such as Russia, in a much better situation than others. Indeed Russia has the largest share of the region's energy resources. Parts of the region (mainly its sea sections) are disputed, pitting concerned littoral countries against each other, e.g., parts of the Barents Sea over which both Norway and Russia claim ownership. Undoubtedly, the five littoral states are very much interested in the prospect of carrying out part of their intercontinental trade through the Arctic sea routes, if they become a reality. By significantly decreasing the shipping costs of their trade, these routes are certainly too tempting to be ignored, especially because all of them have been hard hit by the last decade's financial/economic crises, continuing to this date . The importance of the routes is also linked to a negative economic trend, which started most notably in the 1990s as a result of the littoral states' loss of their economic strength to the emerging economies of Asia and Latin America. For Russia, the Soviet Union's fall and subsequent loss of its economic strength and also its secured markets of the Soviet block and its non-communist allies are yet to be compensated. Hence, lower export costs for Russian products via the Northern Sea Route would certainly help Russia boost its exports and profit margins. Concern about the negative impact of global warming on the Arctic, particularly the rising sea levels that could affect the littoral states, is a major reason for Russia's continued interest in the Arctic.

Apart from the littoral states, the countries of the Greater Arctic region which share parts of the regional areas and resources, form the second tier of the regional stake holders. These are Iceland, Sweden, and Finland, which have access to much smaller parts of the region due to their geographical locations. They are not in a position to lay claim to the bulk of the region — disputed and/or uncharted areas — partly due to their geographical realities leaving no possibility for claiming large shares of the regional territories, waters and mineral and energy resources. In part, their economic, military, and political weaknesses compared to most of the five littoral states make a successful claim to the regional resources beyond their national territories and EEZs rather difficult, at least in the foreseeable future. Yet, these countries surely have other interests in the region, including the use of the Arctic sea routes if they become a reality. As trading countries, reductions in the costs of their imported and exported products will be good news for their small (Iceland) and comparatively small (Sweden and Finland) economies, which are in need of a boost, to offset their losing ground to the emerging Asian and Latin American economies and the hard-hitting financial/economic crises over the last decade. To this, their environmental stake in the Arctic as regional states along the lines mentioned for the five littoral states should be added.

Furthermore, given that major parts of the Arctic Region do not fall within any group of countries' territories and EEZs, they surely do not belong *only* to the eight mentioned countries. More important than this, as mentioned earlier, the region's environmental health is essential for the entire planet's environmental health. Hence, in reality, all countries can be considered stakeholders in the region. However, there is only a small number of non-regional countries capable of entering the Arctic as active players. They are concerned not just about its melting ice, but also the potential of the region's minerals and oil and gas reserves. These countries are mainly China, India, Japan, and South Korea, which, to varying degrees, have established footholds in the Arctic, primarily up to this point, for scientific/environmental purposes. However, as major global energy consumers, they certainly have an eye on the vast regional oil and gas reserves which could potentially address some of their large and, in most cases, growing energy requirements. As it stands, none of these states has a declared plan to explore oil and gas in the Arctic, but all are interested in supporting such exploration through some form of engagement, including investment, in the regional countries' oil/gas development projects. The same type of involvement is also conceivable in the region's mining activities, of course to a much lesser extent given that such engagement will not be crucial for their economies, at least in the foreseeable future. As major trading countries involved in intercontinental trade, these countries are certainly interested in the prospect of conducting a part of their trade via the Arctic sea routes, if and when they are available. Finally, they all share environmental concerns, particularly but not exclusively, the prospect of rising sea levels, as such development will negatively affect their countries' large coastlines.

1.3. ISSUES OF CONCERN

The melting of the Arctic ice is a matter of concern for various reasons. First and foremost and unsurprisingly, the single most important concern is the environment. The melting will have major negative environmental implications affecting the sustainability of life in all its forms, regionally and globally. The disappearance of the Arctic, partially or completely, will further worsen global warming while leading to rising sea levels, affecting all countries with a coastline and particularly the island states, which could even experience total submersion depending on their locations and geological formations.

Second, the dramatic climatic change in the Arctic will encourage the exploitation of the regional energy and mineral resources. The economic and financial importance of these resources, particularly oil and gas, being two types of fossil energy in demand globally, will certainly lead to competition with predictable results. In their attempts to secure large shares of the regional resources, disputes, tensions, and conflicts will arise between and among the eight states in the region, and the five littoral states, in particular. The direct, but most likely, indirect involvement of the non-regional states with an ability to penetrate the region (namely, China, India, South Korea, and Japan) will further intensify the competition and affect regional and global relations.

Third, the possible opening of Arctic sea routes in a few decades is yet another source of concern. Contingent on their sustained availability, use of these passages, even for a few months a year, will substantially reduce the costs of exports for the large trading nations with access to and ability to use these sea routes, providing an obvious economic benefit, as discussed earlier. These exporting nations can potentially increase their exports by pricing their products more competitively due to their lower shipping costs, and thereby expand their share in global markets. For that matter, the issue of free and uninterrupted access to these routes will possibly trigger competition over controlling them, pitting all the regional and non-regional states with large and growing export capabilities and having suitable sea vessels, against each other. Over time, disputes over access to and use of these routes could well turn into inter-state and, quite possibly, inter-regional conflicts affecting regional and international relations.

Finally, military control and an arms race will likely be a by-product of the Arctic's melting ice. The region was militarized during the Cold War by the two major protagonists of the era, the United States and the Soviet Union, each leading a block of like-minded states and corresponding military alliances. Providing a connecting route between the Atlantic and Pacific Oceans, the waters were regularly frequented by their nuclear submarines. The significant American and Soviet military presence was supplemented by other regional countries as part of their defence policies. When the USSR collapsed in 1991, and the Cold War ended, the military significance of the Arctic declined, though all the regional states have maintained some degree of military presence. Especially since the middle of the last decade, the possibility of oil, gas, and mineral extraction and shorter sea routes has rekindled interest in the Arctic. This development has justified a growing military presence, especially but not exclusively, by Russia and the United States and/or a reorganization of the regional military forces to adapt themselves to the new realities in order to protect their respective national interests. Towards this end, the activities of the United States and Russia, as two military giants have been especially significant compared to those of the other regional countries. The United States established the Thule Air Base in Northern Greenland after World War II. On 7 April 2011, changes made to the U.S. military's Unified Command Plan shifted geographic boundaries and revealed the growing importance of the Arctic.[20] The biggest change to the plan assigns U.S. Northern Command (USNORTHCOM) responsibility for the North Pole and its surroundings. USNORTHCOM and US European Command will share responsibility for the Arctic. Russia's significant military presence in the Arctic started in 1937 when the Soviet Union stationed the Red Banner Northern Fleet in Severomorsk with access to the Barents and Norwegian Seas, and the Arctic and Atlantic Oceans; it is responsible for the defence of Northwestern Russia.[21] In March 2009, Russia's National Security Council made public its plans to create a new Arctic military force to defend the country's interests in the disputed polar region.[22]

As small players in the military realm, the other regional countries' recent activities have been less impressive, reflecting their small military capabilities. By and large, they have maintained their limited regional military presence in the region, that started mainly in the post-WWII era. In the late 1950s, Canada set up the Canadian Forces Station Alert on the Northern tip of Ellesmere Island to collect signal intelligence. It also created a permanent military command in Yellowknife, Northwest Territories in 1970, with a view to treat the North as a single theatre of operations; today the country's military responsibility for the North falls under Canada Command created on 1 February 2006.[23] In August 2009, Norway moved its centre of military operations from Jåttå to Reitan located outside of Bodø, thus "becoming the first Arctic state to headquarter its operations command in the High North."[24] Bodø is currently home to the country's main air force base, which deploys aircraft to monitor Russian jet exercises.[25] Finally, Denmark's military interest in the Arctic is evident in its 2011 document entitled, "*Kingdom of Denmark Strategy for the Arctic 2011–2020* ". This document has a stronger focus on the tasks of the Danish Armed Forces in the Arctic providing for them to be visibly present in and around Greenland and the Faroe Islands to enforce sovereignty and surveillance.[26] Along this line, its Armed Forces North Atlantic command structure will be streamlined by the establishment of an Arctic Response Force.[27]

1.4. SOVEREIGNTY DISPUTES

The Arctic is a vast area (21,000,000 sq km) consisting of the Arctic Ocean covered by thick ice and surrounded by permafrost. Parts of the surrounding area are within the territorial limits of the eight Arctic countries. No country, whether regional or non-regional, owns the Arctic Ocean while the regional countries bordering the ocean, including Canada, Denmark, Norway, Russia and the United States, each have 200 nautical miles of EEZ around their Arctic Ocean coasts. Upon ratification of the *United Nations Convention of the Law of the Sea*,[28] all the states falling within this category can make claims to extend their zone within ten years. Pursued by most of these countries, such claims could and probably would create disputes and conflicts over the ownership of certain parts to which more than one regional country has laid claim. This scenario would be even more likely in the areas potentially rich in mineral and energy resources, which would be the main reason for a country to claim them in the first place. Hence, the efforts by the regional countries to extend their ownership of the Arctic region will likely be a trigger for disputes between and among the regional states. Other potential contenders, which are essentially confined to only those non-regional states with a capability for a long-term engagement in the Arctic, namely China, Japan, South Korea, India and possibly Australia, would not have any strong grounds for claiming ownership of any part of the region as they lack any geographical connection to it. Given this reality, the only potentially conceivable scenario could be their laying such claim as part of a global claim of ownership over the entire Arctic Ocean on environmental grounds, i.e., the necessity for global governance of the region to ensure its survival as well as the long-term survival of Planet Earth.

1.5. POTENTIAL POLITICAL/MILITARY CONFLICTS

There is potential for political conflicts between and among the regional states. The ownership of the region's riches, including minerals, especially oil and gas, could trigger such conflicts, adding to other potential triggers such as fishery rights. However, such conflicts are not inevitable as the region seems to have large enough resources to provide a decent share to all the Arctic states. Nonetheless, developing the region's resources is not and will not be an easy undertaking due to the hostile climatic conditions (extreme cold) and the geographically difficult location of many of the known oil and gas reserves, i.e., deep sea fields. These difficulties will become slightly less challenging in the foreseeable future if the current melting process continues due to global warming. Nevertheless, the severe cold, the technological challenges of extracting and exporting the resources, particularly oil and gas, and the resulting high operation costs could well reduce the attractiveness of the resources to the regional states and dampen inter-state ownership disputes. Furthermore, any successful bid to secure ownership over disputed areas would require a military capability to back the claim if necessary. In this regard, Russia and the United States are the only qualified states. Given the various economic/financial difficulties that the two states are currently facing and will continue to experience in the predictable future, they will not likely be keen to escalate any potential ownership dispute into a major political conflict, let alone a military one. Yet, as in any other part of the world, exploiting the potentially rich resources could well instigate conflicts mainly between and among the regional countries. The extent and outcome of these will be determined by the relative importance of the disputed areas to the claimants, their political, economic and military capabilities to back their claims, and the overall regional and international modes which could favour or disfavour such claims.

1.6. ISSUES COVERED IN THE BOOK

Against this background, this book aims to shed light on the main issues pertaining to the Arctic of relevance to its countries and also the rest of the world. As more and more of the ice in the Arctic melts, the main driving force behind this endeavour is, on the one hand, to ponder the survival of this planet, and on the other, to study the energy, environmental, economic, political and military/security consequences arising from resource extraction and export, and also the opening of new trade routes.

At the time of writing, there were two main views on the consequences of the melting ice and its root causes. These are covered in the two chapters of Section I. In Chapter 2, Shawn J. Marshall discusses the human-made nature of this warming, i.e., global warming caused by certain human activities, which have released GHG, especially CO_2, over the last two centuries. Basing his argument on scientific evidence gathered over time, which reflects a steady pattern of Arctic ice shrinkage, his view mirrors that of the majority of scientists specializing in climatic changes, particularly with regard to the status of the sea-based and land-based concentrated ice in the Arctic as well as the glaciers throughout the world. After offering a comprehensive account of this view, he examines various scenarios of the consequences resulting from the Earth's losing a major cooling structure and the rising sea levels, affecting all countries having coastlines along open seas.

A totally different argument is presented by Vyacheslav Mikhailovich Makeev in Chapter 3. While acknowledging that human-generated pollution in different forms has been a factor in damaging the Arctic's fragile ecosystems, he argues that cyclic climatic changes, which have occurred over hundreds of thousands, if not millions, of years are to blame. Using scientific data, he explains that our planet has experienced regular periods of global warming and global cooling as part of a natural cycle caused by both natural terrestrial and celestial forces. Thus, human-made emissions of GHG, while destructive, have not been the main culprit, and human-made efforts to stop the warming cannot significantly change the situation in the Arctic for the better. He notes that yes, certain parts of the Arctic have indeed been melting, but at the same time, other areas have been experiencing cooling not melting.

Should the current melting processes continue, all countries will become increasingly interested in extensive exploitation of the regional assets, namely, oil, gas and mineral extraction and export, and large-scale inter-continental trade for a few months a year through the newly opened Arctic sea routes. Given their natural and direct access to the region, the first natural beneficiaries of this situation are the Arctic states. However, these states have differing stakes based on their currently available opportunities, needs, and capabilities.

To assess these issues, Section II focuses on two Arctic littoral states with different experiences and capabilities to engage in the region. Chapter 4 is dedicated to Russia's approach to and interests in the Arctic. Nadezhda Klimovna Kharlampyeva provides historical background on the significance of the Arctic to Russia and its engagement in the region since the Soviet era. Against this background, and detailing the huge energy, mineral and natural resources in the Russian zone to be developed, she discusses different components of the Russian policy towards the Arctic based on its multi-dimensional interests in the region. Apart from its economic and energy interests, it also has military/security interests as well. Moscow views the Arctic as a major theatre for defending parts of Russia and securing its access to the open seas. While acknowledging the potential for disputes between and among the regional countries, she suggests that the seeds of cooperation do exist, thanks to the membership of the Arctic countries in various regional organizations, which can potentially be used for encouraging regional cooperation.

Chapter 5 offers an account of Norway's interests in and concerns about the Arctic. Arild Moe focuses on the potential sources of conflict in Norway's ties with the region's other states, particularly Russia. He identifies oil as being a commodity which could lead either to cooperation or conflict. Given that the Barents Sea provides access to the region's energy resources, he elaborates on the delimitation agreement, signed in 2010, between Russia and Norway, as a key issue for peace and stability. Yet, the issue of jurisdiction and delimitation of the Arctic continental shelf whose full energy reserves are yet to be proven, is a potential source of dispute in the region for its obvious importance to the two countries as well as the other regional states. Moe also discusses how regional and international cooperation in dealing with the Arctic's environmental challenges is a necessity.

Although the Arctic states are surely the major players in the region, there are many non-regional states (e.g., India as an emerging economy, and Japan as a mature economy) and regional groupings, i.e., the European Union (EU) consisting of many industrialized European countries, with vested interests in the Arctic and the capability to engage there. India, Japan and the EU are distinct from other non-regional states in that they have had a limited presence in the Arctic in the form of scientific expeditions for differing periods of time. Hence, Section III sheds light on their interests and objectives in the Arctic. In Chapter 6, Neil Gadihoke provides an account of the environmental degradation of the Arctic and the melting of its ice and highlights the positive and negative implications for India. He concentrates on the Arctic's energy significance for India as an expanding large economy with a growing energy demand. He also elaborates on the trade importance of the sea routes via the Arctic for India whose international trade has been growing and thus can surely benefit from a shorter sea route to decrease the costs of its exported products. However, these positive implications aside, he also explains India's worry about rising sea levels caused by the melting of the Arctic's ice. The country's long coastline and adjunct areas will be severely affected.

Chapter 7 examines Japan's interests in the Arctic. Hiroshi Ohta spells out the various implications of the Arctic's melting ice for Japan and other regional and non-regional countries. What makes his account different from others is his use of a framework for analyzing the latter, shaped by energy security and environmental security as the two major concerns of Japan when it comes to the Arctic. Referring to the limits of the traditional understanding of security, he tackles these issues within the Japanese context of security, i.e., comprehensive security, which includes non-traditional issues such as environmental security. He also elaborates on Japan's specific reasons for engagement in the Arctic. They include its environmental security concerns, comparable to those of all other states, as well as its energy security concerns, especially in the post-Fukushima era when many Japanese are very worried about the country's continued use of nuclear power.

Chapter 8 discusses the EU's reasons for engagement in the Arctic. With few Arctic members (Denmark through Greenland, Finland, and Sweden), the EU is certainly interested in the region, but, Danila Bochkarev argues that its interests go beyond protecting its members' stakes in the *High North* as the Arctic melting ice could provide opportunities as well as challenges for this regional grouping as a whole. He identifies Brussels' key issue areas as regional governance, climate change (e.g., rising sea levels) and access to the regional bio-marine resources, including minerals (rare earth and precious metals), oil, gas and fish stocks. He also offers an account of the importance of the Northern Sea Route via the Arctic for the EU, tied to its membership's interests in international trade, which could benefit from a shorter and therefore less expensive northern sea passage for such trade compared to the currently used sea route. While the EU is very much interested in the Arctic, he concludes that the EU's role as an "agenda-setter" in the Arctic will remain limited in the near future unless Greenland comes back to it.

Finally, Chapter 9 concludes with Hooman Peimani summarizing the major issues raised in the previous chapters. Against this background, he draws the readers' attention to the major challenges and opportunities arising from the melting of the Arctic ice while identifying the significant trends, which could shape the pace of events in the Arctic and beyond.

BIBLIOGRAPHY

Anisimov, OA, DG Vaughan, TV Callaghan, C Furgal, H Marchant, TD Prowse, H Vilhj'almsson and JE Walsh (2007). Polar regions (Arctic and Antarctic). In *Climate Change 2007: Impacts, Adaptation and Vulnerability. Contribution of Working Group II to the Fourth Assessment Report of the Intergovernmental Panel on Climate Change.* ML Parry, OF Canziani, JP Palutikof, PJ van der Linden and CE Hanson (eds.), 653–685. Cambridge: Cambridge University Press. http://www.ipcc.ch/pdf/assessment-report/ar4/wg2/ar4-wg2-chapter15.pdf [accessed 1/11/2011].

Baffin Island Iron (2008). *Arctic Economics.* http://benmuse.typepad.com/arctic economics/2008/07/baffinlands.html [accessed 1/11/2011].

Baffinland Iron Mines Corporation (2007). *The Mary River Project.* http://www.baffinland.com/MaryRiverProject/default.aspx [accessed 30/10/2011].

Bennett, M (2009). Nordic military developments. *Foreign Policy Association*, 9 June. http://foreignpolicyblogs.com/2009/06/09/nordimilitary-developments/ [accessed 30/10/2011].

Bo'e, J, A Hall and X Qu (2009). September sea-ice cover in the Arctic ocean projected to vanish by 2100. *Natural Geoscience*, 2, 341–343.

Burakova, I (2005). Development of Arctic areas to bring trillions dollars of profit to Russia. *Pravda*, 21 April.

Garamone, J (2011). Unified command plan reflects Arctic's importance. *American Forces Press Service*, U.S. Department of Defense. http://www.defense.gov/news/newsarticle.aspx?id=63467 [accessed 5/11/2011].

Gethin, PAB (1959). Iron ore in Arctic Europe. Polar record. Cambridge Journal [Online]. http://journals.cambridge.org/action/displayAbstract?fromPage=online&aid=5658876 [accessed 5/11/2011].

Glasby, GP and YL Votekhovsky (2009). Arctic Russia: Minerals and mineral resources. *Geochemical Society*. Retrieved from http://www.geochemsoc.org/publications/geochemicalnews/gn140jul09/arcticrussiamineralsandmin.htm#, http://www.arctic.ru/natural-resources/other-minerals [accessed 30/10/2011].

Grim new prediction: Arctic Sea disappear by 2015 (2011). *Alaska Dispatch*, 20 November. http://www.alaskadispatch.com/article/grim-new-prediction-arctic-sea-ice-could-disappear-2015 [accessed 30/10/2011].

Handwerk, B (2004). Arctic melting fast: May swamp U.S. coasts by 2099. National Geographic, 9 November. http://news.nationalgeographic.com/news/2004/11/1109 041109 polar ice.html [accessed 25/10/2011].

Hodge, N (2009). Russia's new Arctic fighting force. Wired, 27 March. http://www.wired.com/dangerroom/2009/03/russias-new/ [accessed 25/10/2011].

Kingdom of Denmark: Strategy for the Arctic 2011–2020. NANOQ. http://uk.nanoq.gl/~/media/29cf0c2543b344ed901646a228c5bee8.ashx [accessed 25/10/2011].

Knappenberger, C (2009). Waxman-Markey: Inconsequential for sea-level rise, too. *MasterResource*, 30 July. http://www.masterresource.org/2009/07/waxman-markey-inconsequential-for-sea-level-rise-too/ [accessed 3/11/2011].

Northern Sea Route (2011). *Voice of Russia*, 22 November. http://english.ruvr.ru/2011/11/22/60807981.html [accessed 3/11/2011].

Northern Fleet. http://en.wikipedia.org/wiki/Northern Fleet [accessed 4/10/2011].

Northern Operations (2011). *The Sentinel*, 1(2). http://www.canadacom.forces.gc.ca/sentinel-sentinelle/vol-1-2/pg6-eng.asp [accessed 5/11/2011].

Robertson, J and B Pierce (2008). 90 billion barrels of oil and 1,670 trillion cubic feet of natural gas assessed in the Arctic. *U.S. Geological Survey*, 23 July. http://www.usgs.gov/newsroom/article.asp?ID=1980&from=rss home [accessed 28/10/2011].

S. Rajaratnam School of International Studies (RSIS) (2011). The Northern Sea Route: Is it really ice-free? Analysis. *Eurasiareview*, 19 August. http://www.eurasiareview.com/19082011-the-northern-sea-route-is-itreally-ice-free-analysis/ [accessed 28/10/2011].

Ships cross Arctic passage in milestone (2009). Associated Press, 9 November. http://www.msnbc.msn.com/id/32800658/ns/usnews-environment/t/ships-cross-arctic-passage-milestone/ [accessed 28/10/2011].

Szumigala, DJ, LA Harbo and RA Hughes (2009). *Alaska's Mineral Industry 2009.* Department of Natural Resources, Division of Geological & Geophysical Surveys. http://serac.dnr.state.ak.us/webpubs/dggs/sr/text/sr064.PDF [accessed 28/10/2011].

United Nations. *United Nation Convention on the Law of the Sea of 10 December 1982.* http://www.un.org/depts/los/convention agreements/texts/unclos/unclos e.pdf [accessed 25/10/2011].

ENDNOTES

1 Emissions scenario prepared by IPCC, a balanced emphasis on energy sources.

2 Emissions scenario prepared by IPCC, an emphasis on fossil-fuels (Fossil Intensive).

3 Knappenberger, C (2009). Waxman-Markey: Inconsequential for sea-level rise, too. *MasterResource*, 30 July. http://www.masterresource.org/2009/07/waxman-markey-inconsequential-for-sea-level-rise-too/ [accessed 3/11/2011].

4 Bo'e, J, A Hall and X Qu (2009). September sea-ice cover in the Arctic ocean projected to vanish by 2100. *Natural Geoscience*, 2, 341–343.

5 Anisimov, OA, DG Vaughan, TV Callaghan, C Furgal, H Marchant, TD Prowse, H Vilhj'almsson and JE Walsh (2007). Polar regions (Arctic and Antarctic). In *Climate Change 2007: Impacts, Adaptation and Vulnerability. Contribution of Working Group II to the Fourth Assessment Report of the Intergovernmental Panel on Climate Change*, ML Parry, OF Canziani, JP Palutikof, PJ van der Linden and CE Hanson, (eds.), pp. 653–685. Cambridge: Cambridge University Press. http://www.ipcc.ch/pdf/assessment-report/ar4/wg2/ar4wg2-chapter15.pdf [accessed 1/11/2011].

6 Handwerk, B (2004). Arctic melting fast: May swamp US coasts by 2099. *National Geographic*, 9 November. http://news.nationalgeographic.com/news/2004/11/1109 041109 polar ice.html [accessed 25/10/2011].

7 Robertson, J and B Pierce (2008). 90 billion barrels of oil and 1,670 trillion cubic feet of natural gas assessed in the Arctic. US Geological Survey, 23 July. http://www.usgs.gov/newsroom/ article.asp?ID=1980&from=rss home [accessed 28/10/2011].

8 *Ibid.*

9 Glasby, GP and YL Votekhovsky (2009). Arctic Russia: Minerals and mineral resources. *Geochemical Society*, July 2009. Retrieved from http://www.geochemsoc.org/publications/geochemicalnews/gn140jul09/arcticrussiamineralsandmin.htm#, http://www.arctic.ru/natural-resources/other-minerals [accessed 30/10/2011].

10 Burakova, I (2005). Development of Arctic areas to bring trillions dollars of profit to Russia. *Pravda*, 21 April.

11 Szumigala, DJ, LA Harbo and RA Hughes (2009). *Alaska's Mineral Industry 2009*. Department of Natural Resources, Division of Geological & Geophysical Surveys, 2009. http://serac.dnr.state.ak.us/webpubs/dggs/sr/text/sr064.PDF [accessed 28/10/2011].

12 Baffinland Iron Mines Corporation (2007). *The Mary River Project*. http://www.baffinland.com/MaryRiverProject/default.aspx [accessed 30/10/2011].

13 Baffin Island Iron (2008). *Arctic Economics*. http://benmuse.typepad.com/arcticeconomics/2008/07/baffinlands.html [accessed 1/11/2011].

14 Gethin, PAB. Iron Ore in Arctic Europe, Polar Record 1959. *Cambridge Journal* [Online]. http://journals.cambridge.org/action/displayAbstract?fromPage=online&aid=5658876 [accessed 5/11/2011].

15 *Ibid.*

16 Northern Sea Route (2011). *Voice of Russia*, 22 November. http://english.ruvr.ru/2011/11/22/60807981.html [accessed 3/11/2011].

17 Ship cross Arctic passage in milestone. *Associated Press*, 9 November. http://www.msnbc.msn.com/id/32800658/ns/us news-environment/t/ships-cross-arctic-passage-milestone/ [accessed 28/10/2011].

18 S. Rajaratnam School of International Studies (RSIS) (2011). The Northern Sea Route: Is it really ice-free? Analysis. *Eurasiareview*, 19 August. http://www.eurasiareview.com/19082011-the-northern-searoute-is-it-really- ice-free-analysis/ [accessed 28/12/2011].

19 Grim new prediction: Arctic sea disappear by 2015 (2011). *Alaska Dispatch*, 20 November. http://www.alaskadispatch.com/article/grim-new-prediction-arctic-sea-ice-could-disappear-2015 [accessed 30/10/2011].

20 Garamone, J (2011). Unified command plan reflects Arctic's importance. *American Forces Pres Service*, US Department of Defense. http://www.defense.gov/news/newsarticle.aspx?id=63467 [accessed 5/11/2011].

21 Northern Fleet. http://en.wikipedia.org/wiki/Northern Fleet [accessed 4/10/2011].

22 Hodge, N (2009). Russia's new Arctic fighting force. *Wired*, 27 March. http://www.wired.com/dangerroom/2009/03/russias-new/ [accessed 25/10/2011].

[23] Northern Operations (2011). *The Sentinel*, National Defence and the Canadian Forces, 1(2). http://www.canadacom.forces.gc.ca/sentinel-sentinelle/vol-1-2/pg6-eng.asp [accessed 5/10/2011].

[24] Bennett, M (2009). Nordic military developments. *Foreign Policy Association*, 9 June. http://foreignpolicyblogs.com/2009/06/09/nordicmilitary-developments/ [accessed 30/10/2011].

[25] *Ibid.*

[26] Kingdom of Denmark: Strategy for the Arctic 2011–2020. NANOQ. http://uk.nanoq.gl/~/media/29cf0c2543b344ed901646a228c5bee8.ashx [accessed 25/10/2011].

[27] *Ibid.*

[28] United Nations. *United Nations Convention on the Law of the Sea of 10 December 1982*. http://www.un.org/depts/los/convention agreements/texts/unclos/unclos e.pdf [accessed 25/10/2011].

SECTION I

IS THE ARCTIC MELTING?

2

FROM WHITE TO BLUE: THE SHRINKING ARCTIC CRYOSPHERE

Shawn J. Marshall

2.1. INTRODUCTION

The global cryosphere encompasses the world's snow and ice, including glaciers and ice sheets, sea ice, lake and river ice, permafrost, seasonal snow, and ice crystals in the atmosphere. Figure 2.1 illustrates the typical winter extent of global snow and ice cover, based on mean observed snow and ice area. The global extent of snow and ice is compiled in Table 2.1. The Northern Hemisphere has a proportionally greater high-latitude land mass, giving a distinctly Northern flavor to the seasonal cryosphere.

The distribution of the continents influences several other aspects of cryospheric sensitivity to climate change in the Arctic region. The land-sea configuration promotes meridional mixing of air and water masses, which transports heat effectively from mid-latitudes rendering the Arctic susceptible to the marked Northern Hemisphere warming in recent decades. Unlike Antarctica, most of the high Arctic regions are only thinly veiled in snow and ice and are at a low elevation (hence, warmer — indeed, close to the melting point of ice). The sea ice cover which blankets the Arctic Ocean and sub-polar seas is only decimeters to meters thick, with 65% of the ice area melting away each summer.[1] This makes sea ice very responsive to inter-annual and multi-year climatic fluctuations.

Fig. 2.1. Global Snow and Ice Cover in Late Winter in Each Hemisphere.
Source: Image adapted from the UNEP/GRID-Arendal Global Outlook for Ice and Snow (2007). Original cartography by Hugo Ahlenius. http://maps.grida.no/go/graphic/cryosphere-winter-seasons-northern-and-southern-hemispheres [accessed 4 June 2009].

Loss of sea ice has received well-documented positive feedback effects under climate change, including increased absorption of solar radiation (a decline in albedo), warming of the ocean, and increased sensible heat flux from the water to the ice. The associated increase in tropospheric water vapor and cloud cover in the Arctic region may also be negatively impacting sea ice, through increased downward fluxes of infrared radiation. These have contributed to the well-documented reduction in the volume of Arctic ice area in recent years.[2–4] Sea ice loss has been driven by a combination of atmospheric warming and the advection of warm North Atlantic water into the Arctic basin, which melts and thins the sea ice from below. The suite of positive feedback amplifies the climate forcing and creates multi-year memory in ice volume trends.

Coincident with the summer sea ice decline in the Arctic, every other aspect of the Northern cryosphere is also registering a strong response to climate warming. Seasonal snow, permafrost, lake ice, mountain glaciers, and the Greenland Ice Sheet have all been in a

Table 2.1 Area and Volume of the Global Cryosphere.[7] (Sea Ice Data is from 1979–2010 and Northern Hemisphere Snow Data is from 1966–2010).

Perennial Ice	Area (10^6 km^2)	Volume (10^6 km^3)	Sea level equivalent (m)
Greenland Ice Sheet[a]	1.7	2.9	7.1
Antarctic Ice Sheet[b]	13.3	25.4	56.2
Mountain glaciers[c]	1.1	0.22–0.38	0.56–0.97
Permafrost[d]	22.8	0.01–0.04	0.03–0.10

		Area (10^6 km^2)			Extent (10^6 km^2)		
Snow and Sea Ice		Minimum	Maximum	Mean	Minimum	Maximum	Mean
Sea ice[e]:	Northern	4.8	13.6	9.8	6.6	15.5	11.8
	Southern	1.9	14.5	8.7	3.0	18.8	12.0
Snow[f]:	Northern				3.1	46.7	24.9
	Southern				13.9	15.1	14.3

Sources:
[a]Bamber, JL, RL Layberry and SP Gogenini (2001). A new ice thickness and bed data set for the Greenland ice sheet 1: Measurement, data reduction, and errors. *Journal of Geophysical Research*, 106(D24), 33,773–33,780.
[b]Lythe, MB, DG Vaughan and the BEDMAP Consortium (2001). A new ice thickness and subglacial topographic model of Antarctica. *Journal of Geophysical Research*, 106(B6), 11,335–11,351.
[c]Adapted from Radić, V and R Hock (2010). Regional and global volumes of glaciers derived from statistical upscaling of glacier inventory data. *Journal of Geophysical Research*, 115(F01010), doi:10.1029/2009JF001373.
[d]Zhang, T, RG Barry, K Knowles, JA Heginbottom and J Brown (2008). Statistics and characteristics of permafrost and ground ice distribution in the Northern Hemisphere. *Polar Geography*, 31(1–2), 47–68.
[e]UNEP/GRID-Arendal (2007). Mean snow-cover extent in the Northern Hemisphere 1966–2006. UNEP/GRID-Arendal Maps and Graphics Library. Available at: http://maps.grida.no/go/graphic/mean-snow-cover-extent-in-the-northern-hemisphere-1966–2006. Updated to 2010 with data from the Rutgers University Global Snow Lab, http://climate.rutgers.edu/snowcover [accessed 7 April 2011].
[f]Fetterer, F, K Knowles, W Meier and M Savoie (2002, updated 2011). *Sea Ice Index*. Boulder, CO: National Snow and Ice Data Center. Digital media, http://nsidc.org/data/seaice_index/index.html. [accessed 7 April 2011].

state of decline over recent decades.[5,6] Glacier and ice sheet retreat are of particular concern, as the transfer of ice from the land to the ocean has become the leading driver of global sea level rise over the past 20 years. This is due to a combination of increased melting of mountain glaciers in the 1990s and 2000s,[8] dramatic increases in melting and runoff in Greenland,[9,10] and accelerations in major outlet glaciers in both Greenland and West Antarctica, which transfer ice directly to the ocean via iceberg calving.[11,12] There is a great deal of uncertainty in projections for future sea level rise, because it is unclear whether the Greenland and Antarctic Ice Sheets will continue to lose mass at an accelerated rate. However, it is clear that the polar ice sheets are more sensitive to climate warming than previously thought. This means that projected global and high-latitude warming is likely to increase the rate of sea level rise over the coming decades.

This is crucial because sea level rise has the potential to be one of the most disruptive and damaging impacts of global climate change. The World Bank estimates that a 1-meter sea level rise would displace at least 56 million people worldwide,[13] with increased flood risk to over 200 million people. The direct impacts will exceed this, even with a sea level rise of tens of centimeters, as 150 of the world's 197 sovereign nations (76.1%) border on the ocean. This includes 664 million people living in the world's 98 largest coastal cities: urban centers with a direct connection with the ocean;[14] and many more millions living in smaller coastal settlements. The economic costs of buttressing coastal infrastructure against rising waters could be severe, and may be untenable in much of the developing world.

This review focuses on three main aspects of climate and cryospheric change in the Arctic region, i.e., reductions in Arctic sea ice, changes in the Greenland Ice Sheet, and implications for global sea level rise. The latter is superimposed on processes that affect regional sea level, which is of significant concern in the tropics. While changes in the Arctic sea ice do not directly influence sea level, they do appear to be affecting the Greenland Ice Sheet. The changing face of the Arctic has many other social, political, and economic impacts. Sea ice decline is raising interest in transportation and economic opportunities, especially oil and gas exploration and mineral extraction, at the same time as it is arousing geopolitical tensions, while the impacts of sea level rise are more global and ultimately more grave, particularly for tropical coastal nations.

In this chapter, the author has summarized the current scientific understanding of what is driving these changes, what can be expected in the coming decades, and what this might mean for climate change adaptation.

2.2. ARCTIC AMPLIFICATION OF CLIMATE CHANGE

Temperature fluctuations in the Arctic are coupled with global mean temperature variability but possess a greater magnitude: a phenomenon known as Arctic amplification (Fig. 2.2). This arises for a number of reasons, mostly associated with climate feedback discussed above. These positive feedback are reasonably well understood and captured by climate models.[15] With warmer temperatures at low latitudes, the advection of moist, warm air into the Arctic region increases. Loss of snow and ice decreases the albedo and increases the area of open water, which has a direct warming influence. Arctic

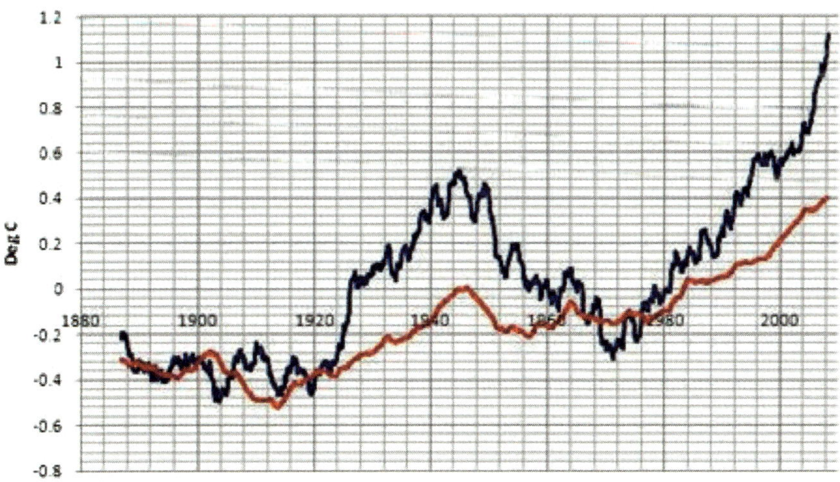

Fig. 2.2. Arctic Amplification of Global Warming, 1880–2008.

Note: The red and blue lines show global average and Arctic (60–90°N) temperatures, respectively.

Source: Monthly data smoothed using a seven-year filter. Data from the ESSTv.3 global reconstruction of the US National Climate Data Center. http://www.ncdc.noaa.gov/oa/climate/research/sst/ersstv3.php [accessed 10 April 2011].

cloud cover tends to have a net warming effect, as annual solar radiation is low at these latitudes (hence infrared warming outweighs the effects of reduced incoming solar radiation).

The extent of Arctic amplification of climate variability is uncertain, but it was a factor of about two in the smoothed 20th century temperature record of Fig. 2.2. The Arctic has warmed by approximately 1.2°C since the 1970s, relative to a global average temperature increase of about 0.5°C. This is regionally variable; parts of the Arctic are warming at an even greater rate, and parts of the Arctic are stable or even cooling, depending on the details of the local and regional energy balance. Nor are the processes linear in time, so it is difficult to project whether continued global temperature increases will cause roughly twice as much warming in the Arctic. If so, the projected temperature rise of 2–4°C for this century will mean 4–8°C in the Arctic. This would threaten many aspects of the Northern cryosphere, including sea ice, permafrost, and the Greenland Ice Sheet.

2.3. WHITHER ARCTIC SEA ICE?

Figure 2.3 plots the anomaly in monthly Arctic sea ice extent for the period 1953–2010. This illustrates both the variability in Arctic ice cover and the decline since the early 1970s. Arctic sea ice loss has been greatest in late summer, when it reaches its annual minimum. This record comes from a combination of satellite data, ship records, and historical ice charts. The satellite record of sea-ice extent began in late 1978, with complete Arctic coverage available since that time. Over the period 1979–2000, the average August–September ice area in the Arctic was 5.4 million km^2. This declined to 4.2 million km^2 (78%) for the period 2000–2010 and 3.8 million km^2 (71%) from 2005–2010.[1] The linear trend for September ice area for the period 1979–2010[16] was $-0.071\,\mathrm{km^2yr^{-1}}$ (Fig. 2.4). This gradual "opening of the Arctic" has generated tremendous interest in the possibility of ice-free conditions in the Arctic in the coming decades.

It is a matter of speculation as to how long it will be before ice-free summer conditions become a reality. Climate model projections

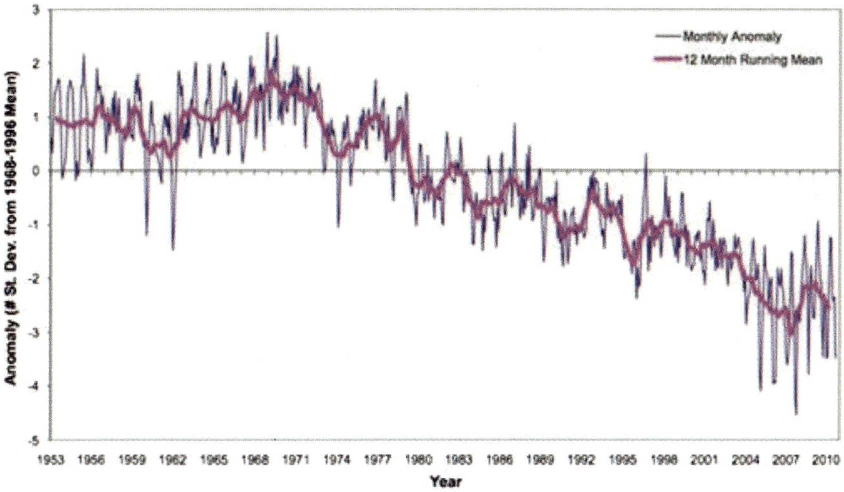

Fig. 2.3. Monthly Sea Ice Extent Anomalies, January 1953 to September 2010.

Source: Data from 1953–1979 is derived from operational ice charts and other sources (UK Hadley Center). Passive microwave satellite data is available from 1979–2010.[1] Figure from W. Meier and J. Stroeve, National Snow and Ice Data Center, University of Colorado, Boulder CO. http://nsidc.org/data/nsidc-0192.html [accessed 12 December 2010].

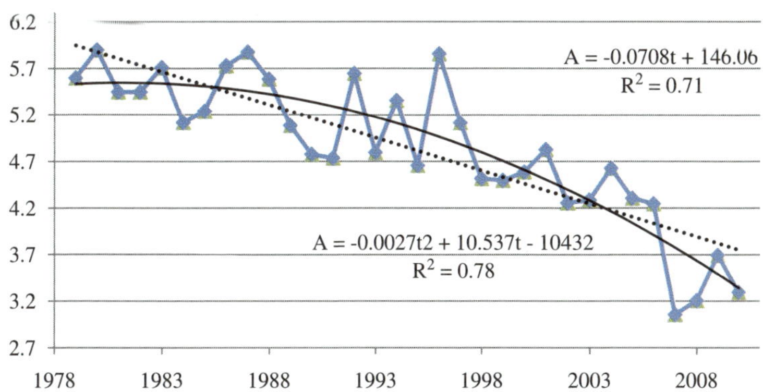

Fig. 2.4. Arctic Sea Ice Area in September (10^6 km^2), 1979–2010.

Note: Linear (dotted) and quadratic (solid) trendlines are shown, along with the goodness-of-fit (R^2) for each.

Source: Figure by author.

indicate that this may occur as early as the 2050s, if global warming continues as anticipated.[17–19] It should be noted that sea ice is poorly represented in many models so there is a lot of uncertainty around these projections. In particular, models did not forecast the recently observed reductions in summer ice extent in the Arctic.[20] Models systematically underestimate recent sea ice decline; it is playing out more quickly than predicted. This implies that the climate models are not capturing all of the relevant positive feedback.

A linear extrapolation of September ice area, based on the observed rate of change of $-0.071\,\text{km}^2\,\text{yr}^{-1}$ (Fig. 2.4), gives an estimate of 53 years (year 2063) before the Arctic is completely ice-free in late summer. Positive feedback mean that the system is nonlinear, and indeed a quadratic fit to the observed sea ice decline fits the data better than a linear relation (Fig. 2.4). Extrapolating from the quadratic fit, September ice cover would disappear by the mid-2030s.

Statistical extrapolation is unreliable because the sea ice system is not that simple and it is very responsive to year-to-year temperature changes. Recent trends are not certain to continue. In addition, multiyear ice that piles up against the coast of the Canadian Arctic Archipelago and Northern Greenland is likely to persist. Ocean gyres and wind cause ice convergence in these regions, where the ice is several meters thick and is likely to survive the summer melt season. On the other hand, the Arctic need not be completely ice free before ship navigation becomes much simpler and safer (e.g., the open water corridors in September 2009 in Fig. 2.5b).

Although seasonally ice-free conditions are likely to be manifest at some point this century, the Arctic Ocean will continue to be ice covered for much of the year for the foreseeable future. The Arctic has warmed, but when the sun sets in autumn for the long, dark winter, temperatures still reach below $-30°\text{C}$ for sustained periods, driving sea ice formation over most of the Arctic (Fig. 2.5a). Although this icepack is thinning and retreating, it must melt away each summer before the Arctic Ocean becomes accessible to all vessels, but the heaviest icebreakers. For many years to come, residual ice from July through September may be a hazard to ship traffic as it can blow into shipping lanes. Hence, there will be ongoing

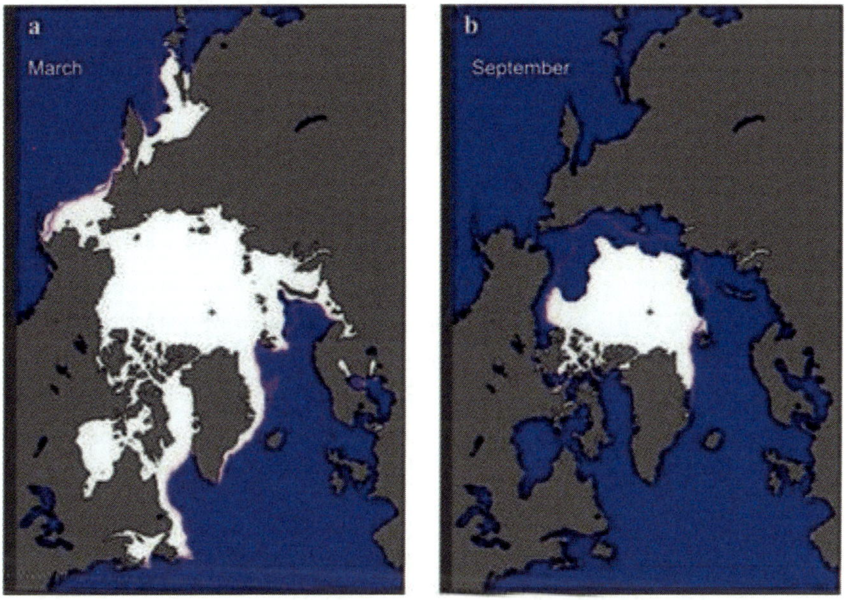

Ice Area: 13.3x10^{6} km^{2} **Ice Area: 3.7x10^{6} km^{2}**

Fig. 2.5. Maximum and Minimum Sea-Ice Extent in the Northern Hemisphere.
Note: Plots show the monthly minimum and maximum ice extents in 2009, along with the median value for the period 1979–2000 (magenta lines).
Source: Images adapted from the US National Snow and Ice Data Center, Boulder, CO, http://nsidc.org/data/g02135.html [accessed 12 November 2009].

hazards in the coastal shipping lanes (Russia's Northern seaway and North America's Northwest Passage). Inter-annual variability will also make these routes unreliable with respect to long-term planning and infrastructure investments. In cold years, heavy ice cover will persist late into the summer, and shifting patterns of ice convergence and divergence will introduce weekly to monthly variability in ease of navigability. Good weekly to seasonal ice forecast skill in the Arctic coastal zones will be essential to Arctic navigation and economic activities.

Despite these concerns, the opening of the Arctic is under way and is expected to continue. In addition to increased open water, the trend towards thinner ice with reduced multiyear pack ice[21,22] reduces shipping hazards and eases the demand on icebreakers. It is probably a question of when, not if, Arctic shipping routes become viable this century.

2.4. DECLINE OF THE GREENLAND ICE SHEET

Arctic sea ice loss does not directly impact global sea levels, as it already displaces sea water while floating in the ocean. In other words, this water is already in the ocean. On the other hand, melting of land ice (e.g., glaciers, ice sheets, and permafrost) does affect sea levels when it runs into the ocean. Sea ice loss also has indirect impacts on terrestrial ice. Warming of the Arctic, which is amplified by reduced sea ice, is having a major impact on one of the main drivers of global sea level rise: the Greenland Ice Sheet. This ice sheet covers more than 80% of Greenland and harbors $2.9 \times 10^6\,\mathrm{km}^3$ of ice, equivalent to 7.2 meters of global eustatic sea level (msl).

Like most of the world's mountain glaciers, the Greenland Ice Sheet is in decline because warmer temperatures are creating a longer and more intense summer melt season. Unlike most mountain glaciers, however, much of the Greenland Ice Sheet is still too cold to experience summer melting; interior regions of the ice sheet have a high enough elevation that summer temperatures remain below 0°C. This actually makes the ice sheet even more vulnerable and nonlinear with respect to its climate sensitivity, because rising temperatures increase the area of the ice sheet that experiences melt (Fig. 2.6). Melting in Greenland is therefore occurring over a larger area of the ice sheet, over a longer melt season, and with more intensity.[23,24]

Some of the snow and ice that melts percolates into the ice sheet and refreezes. However, the majority of it runs off to the ocean. The average annual runoff over the last 50 years is estimated to be between 200–300 gigatons of water per year ($\mathrm{Gt\,yr^{-1}} = 10^{12}\,\mathrm{kg\,yr^{-1}}$).[25] Note that 357 Gt equates to 1.0 mm of global eustatic sea level. Runoff increased by 3% per year over the period 1990–2007[26] and values over the last decade are close to 400 Gt yr^{-1}.[27]

Annual snow accumulation exceeds the runoff in Greenland, with snowfall totals of about 600 Gt yr^{-1}.[28] This indicates that Greenland is still in a state of positive surface mass balance (net annual snow

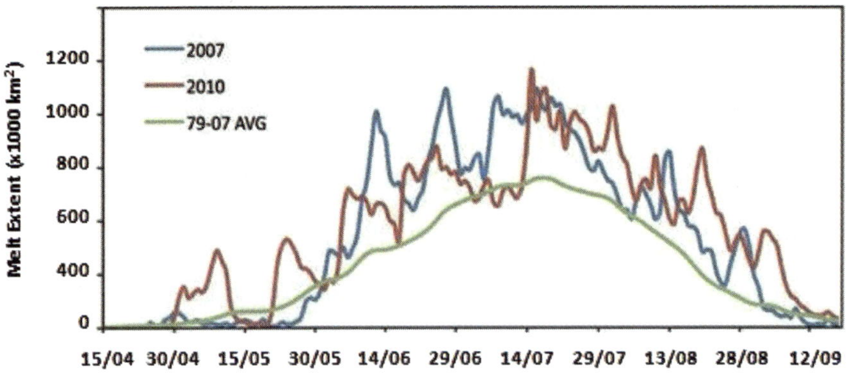

Fig. 2.6. Greenland Ice Sheet Melt Area as Derived from Passive Microwave Remote Sensing.[36,37]

Note: The green line indicates the average area of the ice sheet experiencing melt from 15 April to 15 September over the period 1979–2007. The red and blue lines show the data for 2007 and 2010, respectively.
Source: A new summer melt record was established in 2010. Image from Box, JE, J Cappelen, D Decker, X Fettweis, T Mote, M Tedesco and RSW van de Wal (2010). Greenland, Arctic Report Card 2010. http://www.arctic.noaa.gov/reportcard/greenland.html [accessed 4 May 2011].

accumulation exceeds the runoff). However, 300–400 Gt yr^{-1} of ice is lost through iceberg calving and basal/lateral melting of the ice that is in contact with the ocean.[29,30] Combined with the runoff, the net mass balance is negative in Greenland in recent years, contributing to global sea level rise.

This negative mass balance state has been measured by the GRACE gravity satellite instrument since 2002 (Fig. 2.7).[31] This satellite data does not provide information about glaciological processes (e.g., melting versus calving), but it provides an estimate of the total mass loss from the ice sheet. The mass imbalance increased from −140 Gt yr^{-1} in 2002–2003 to −290 Gt yr^{-1} for 2007–2009.[32] This accelerated ice loss is associated with all three mechanisms of ablation on the ice sheet: increased melt/runoff, increased melt rates at the ice-ocean interface, and speedup of outlet glaciers.[33]

Atmospheric warming is driving higher melt rates on the ice sheet, but the increased mass loss where the ice sheet is in contact with the ocean is not fully understood. Much of this may be due to ocean

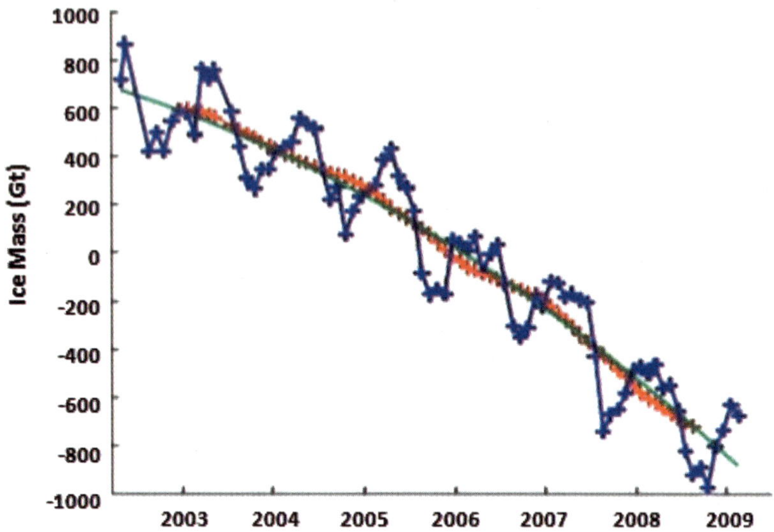

Fig. 2.7. GRACE Gravity Satellite Measurements of Ice Mass Change in Greenland, 2002–2009.

Note: The blue line shows the seasonal cycle of snow accumulation and runoff, based on the monthly GRACE data. The red and green lines indicate the smoothed mean annual data and a quadratic fit to the data, respectively.
Source: Velicogna (2009). Increasing rates of ice mass loss from the *Greenland* and Antarctic ice sheets revealed by GRACE. *Geophysical Research Letters*, 36, L19503, Copyright American Geophysical Union, reproduced by permission.

warming,[34,35] which increases marine melt rates and causes thinning of floating outlet glaciers and ice shelves. This thinning, in turn, leads to drawdown, increased slopes, and increased ice flux from the interior of the ice sheet. It can also precipitate calving events, as thinner, fast-flowing ice is more prone to through-going crevasses and fracture propagation. These processes have multiyear to decadal timescales.

So once a marine outlet glacier begins to retreat, it is apt to continue and propagate inland. Loss of sea ice may play a role in facilitating this process, through resultant ocean warming and also through dynamic feedback where icebergs are more easily evacuated from fjords and they are in open-water conditions. When they are not readily evacuated, icebergs and sea ice can form a stiff, resistant "plug" of ice known as sikkussaq, which provides back-pressure on floating glacier tongues, suppressing calving. Meltwater-driven speedup of ice sheet flow is another relevant process at the ice sheet margins, with a direct sensitivity to climate warming.[38,39] If meltwater penetrates through to the ice sheet base and is not effectively drained from the bed, it can build up to give high basal water pressures, promoting ice-bed decoupling and sliding flow. In this flow regime, glaciers and ice sheets can flow much quicker than they do through purely internal deformation, increasing the flux of ice to the ocean or the ice sheet margin, where it melts at higher rates. This process clearly occurs in Greenland, manifests through summer speedups near the ice sheet margin,[40] but it is not yet clear how much of the recent mass loss in Greenland is associated with this process.

When considered together, direct increases in surface melting/runoff, meltwater lubrication of basal flow, loss of sea ice, and ocean warming, all offer direct coupling between climate warming and ice sheet retreat. This makes the Greenland ice sheet very sensitive to warming in the Arctic region, and it is likely that Greenland will be one of the main contributors to global sea level rise in the coming decades. It is well-established in the paleoclimate record that much of the Greenland Ice Sheet melted away during earlier warm periods in the Earth's history, such as the last interglacial period, 125,000 years ago.[41,42] Temperatures in the Arctic region were 3–5°C warmer than present at that time, similar to what is expected in the Arctic by the middle or end of this century. However, it is not known how quickly the ice sheet retreated during the last interglacial period; it is likely to require several millennia for the entire ice sheet to decay.[43]

Although the observational evidence for glacier and ice sheet decline is mounting, the recent, dramatic changes in ice dynamics cannot yet be simulated by glaciological models. The relevant fast-flow physics and ice-marginal processes are not well understood and are absent or overly simplified in the current generation of models. Similarly, the important three-way interactions among the ice sheet, ocean, and atmosphere are poorly represented in climate models. This is particularly true for ice-ocean coupling, which appears to be driving recent changes in Greenland and Antarctica. Current models, therefore, lack some of the dynamic feedback and processes that are salient to predictions of how rapidly ice sheets can respond to climate forcing.

Bearing in mind the missing ice-dynamical processes and the poor representation of ice sheets in climate models, a number of insights are available from modeling studies. It is clear that the Greenland Ice Sheet is very sensitive to climate warming and that, in a steady state, a regional warming of 4°C or more is likely to be sufficient to precipitate ice sheet retreat in Southern Greenland.[44–46]

Based on an ensemble of models, including high-resolution (ca.1°) climate model representations of temperature and precipitation patterns over the ice sheet, an average annual regional warming of 4.5°C appears capable of pushing the Greenland Ice Sheet into a negative surface mass balance.[47] A warming of 4.5°C in Greenland corresponds to an average global warming of 3.1°C in an ensemble of global climate models.[48] This is an average polar amplification factor of about 1.5 over Greenland, less than the average value for the Arctic. Part of the moderate polar amplification factor here may be due to the cooling influence of weakened deepwater formation in the Nordic Seas that is predicted by many climate models in recent decades. This is not necessarily due to Greenland runoff. The freshwater balance in the North Atlantic and its sub-polar seas is more strongly influenced by variability in sea ice and rainfall.

The critical warming value of 4–5°C for Greenland to go into a state of negative surface mass balance is often quoted as a threshold for "irreversible" ice sheet decline. Dynamical ice losses would place the Greenland Ice Sheet into a negative total mass balance well before this surface mass balance threshold was met — we appear to have reached this threshold in the 1990s and 2000s with a regional warming of close to 1°C. The temperature threshold where surface mass balance becomes negative is considered to be critical to the viability of the ice sheet because thinning provides a positive feedback to ice sheet decay. If this proceeds too far, ice sheet drawdown would make it difficult to reverse the decline, even with temperature stabilization.

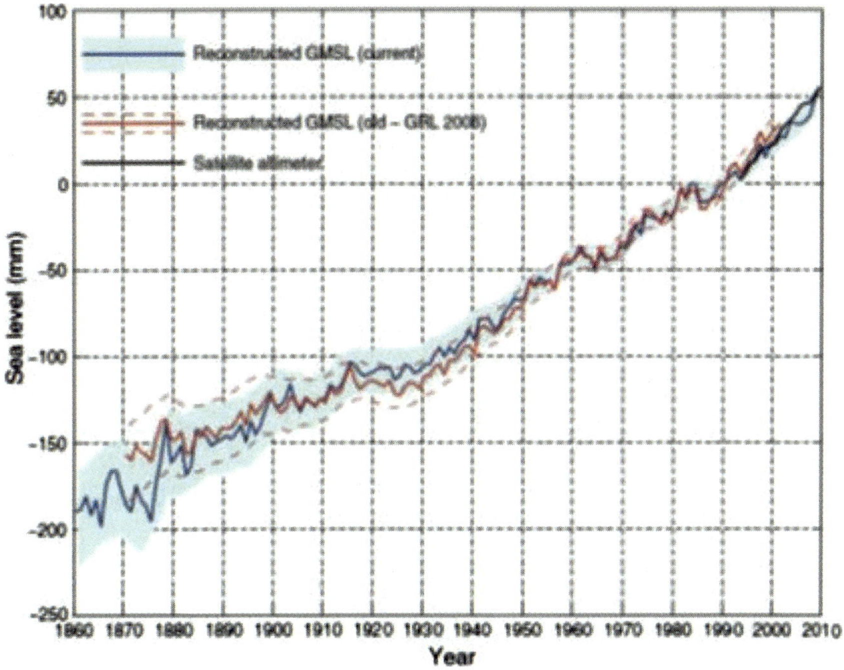

Fig. 2.8. Observed Global Eustatic Sea Level Rise, 1860–2009.

Note: The red and blue lines indicate different reconstructions from the global tide-gauge record.

Source: Church, JA and NJ White (2011). Sea-level rise from the late 19th to the early 21st century. *Surveys in Geophysics*, doi:10.1007/s10712-011-9119-1. Copyright Springer.

2.5. SEA LEVEL RISE

Global mean sea level rose by about 18 cm in the 20th century.[49] Figure 2.8 presents a recent reconstruction based on available tide gauge and satellite sea-surface altimetry records.[50] About two-thirds of the historical sea level rise was due to heat storage and thermal expansion in the upper few 100 m of the oceans. Melting and retreat of mountain glaciers was the other main contributor. As discussed in the above sections, ice loss from the Greenland stepped up in the late 20th century and in the first decade of the 2000s. Similar processes at the ice-ocean interface have driven mass loss in the West Antarctic

Ice Sheet in recent decades.[51,52] Combined with increased rates of mountain glacier retreat, this has driven a rough doubling in the rate of global sea level rise relative to the 20th-century average; the oceans rose at $3.2 \pm 0.4\,\text{mm}\,\text{yr}^{-1}$ for the period 1993–2010.[53] Runoff from glaciers and ice sheets is now roughly equivalent to the sea level rise associated with thermal expansion of the oceans. There is considerable uncertainty regarding future projections of sea level rise. The Intergovernmental Panel on Climate Change (IPCC) projects a sea level rise of 18–59 cm by 2100, with a "best estimate" of 37 cm and an upper bound of 84 cm (Fig. 2.9), with the caveat that this is a

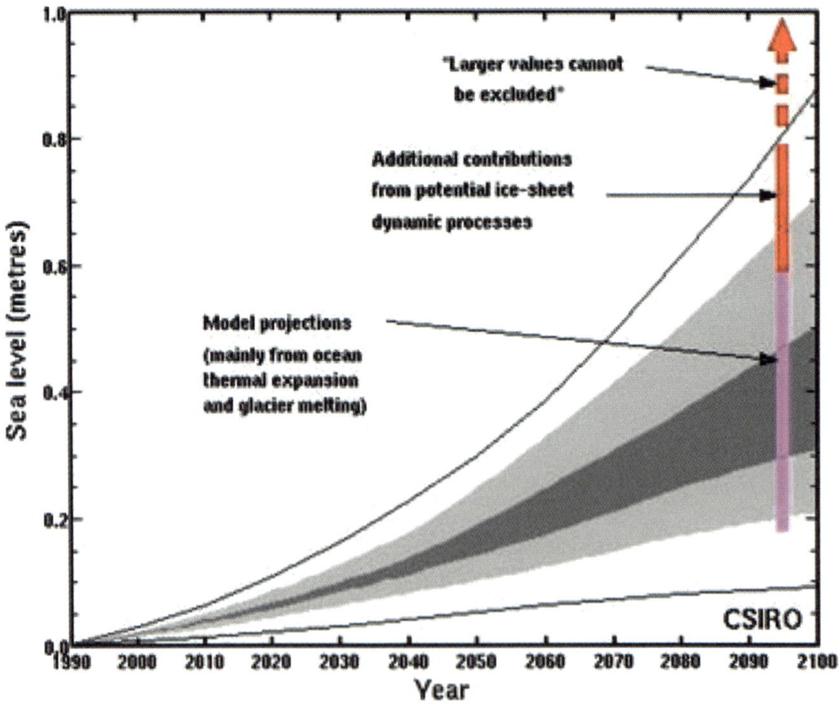

Fig. 2.9. Model Projections of 21st-Century Sea Level Rise.

Note: The black lines and grey shading indicate the IPCC forecast envelope. The red arrows indicate the potential higher contributions associated with ice sheet instabilities.

Source: Commonwealth Scientific and Industrial Research Organization of Australia (2011). http://www.cmar.csiro.au/sealevel/sl proj 21st.html [accessed 10 April 2011].

"model-based range excluding future rapid dynamical changes in ice flow".[54] Limitations in the understanding of glacier dynamics make forecasts difficult.[55,56] The IPCC goes on to note that "Larger values cannot be excluded, but understanding of these effects is too limited to assess their likelihood or provide a best estimate or an upper bound for sea level rise." Because the IPCC's upper bound does not factor in the potential destabilization of major sectors of the Greenland or Antarctic Ice Sheets, it is not an actual upper limit. For a given temperature scenario, sea level rise associated with thermal expansion in the oceans can be estimated reasonably well, as it is expected to be a generally linear process. However, ice sheet instability is nonlinear. This makes it difficult to estimate the most likely ice sheet scenarios, and it also means that uncertainties in the ice sheet contribution to sea level are not Gaussian, but are positively skewed (e.g., Fig. 2.9).

This non-normal uncertainty in projected ice sheet changes has been implicitly cautioned by the IPCC (quoted above), but it has led to misinterpretations about the IPCC sea level projections. While the IPCC numbers were not intended to be applied without taking account of these caveats, there is a tendency for numbers from the tables and graphs to be adopted without regard to the "fine print". As a result, the IPCC sea level forecasts have been challenged as being too conservative.[57–59] As a result, there is a danger that major policy and infrastructure decisions may be made based on sea level rise estimates that are not representative of the most likely or maximum possible scenarios.

This concern is valid since it is rooted in recent observational evidence that the world's glaciers and ice sheets are more vulnerable and reactive than anticipated to climate warming. Indeed, the IPCC's best estimate of 37 cm seems particularly conservative given the current rate of sea level rise of 3.2 .mm yr^{-1}. Extrapolation of this rate of change forward gives approximately 30 cm of sea level rise by 2100. This assumes very little increase in the rate of sea level rise, despite the projected global warming and the expected Arctic amplification of that warming.

While the IPCC forecasts have been criticized, the estimates are based on the best available science. Current models[60,61] are believed to provide good insight into the surface mass balance of the polar ice sheets (snow accumulation and melt) for a particular climate perturbation. However, these models lack the dynamical feedback and processes that are salient to predictions of how rapidly ice sheets can respond to climate forcing. The relevant fast-flow physics and ice-marginal processes are not well understood and are absent or overly simplified in the current generation of models. Similarly, the critical three-way interactions with the ice sheet, ocean, and atmosphere are poorly represented. Models lacking in physics introduce systematic biases that make them conservative. If the ice sheets do not destabilize, the forecasts may be about right; if the recent dynamical changes in the polar ice sheets continue (ice shelf loss; thinning and speedup of marine outlet glaciers), the forecasts may severely underestimate sea level rise.

Since deterministic models of ice sheet dynamics cannot yet address the challenge of forecasting sea level rise, the void in knowledge has been filled by simplistic empirical and heuristic models. Based on the statistical correlation between mean global temperature and rates of sea level rise over the historical period, along with IPCC's temperature projections, the 21st century sea level rise is likely to be about 1 meter.[62] Subsequent refinements and variations on this empirical methodology have reached similar conclusions, with estimates of SLR that range from 0.8 to 1.6 meters by the year 2100.[63,64] Figure 2.10 illustrates these empirical estimates. The envelope of possibility expressed in the blue shading of this figure is associated with different statistical models and different projections of 21st-century global temperature change. These empirical estimates predict sea level rise by 2100 to be about three times higher than the IPCC forecasts.[65]

The causal link between temperature and sea level is direct, as global temperature influences the main factors associated with sea level rise: ocean temperatures (through both thermal expansion and impacts of ocean warming on the ice sheets), glacier and ice sheet melt, and ice shelf instability. In some locations, warming also appears to influence rates of ice sheet flow through meltwater penetration to the bed and enhanced basal sliding.

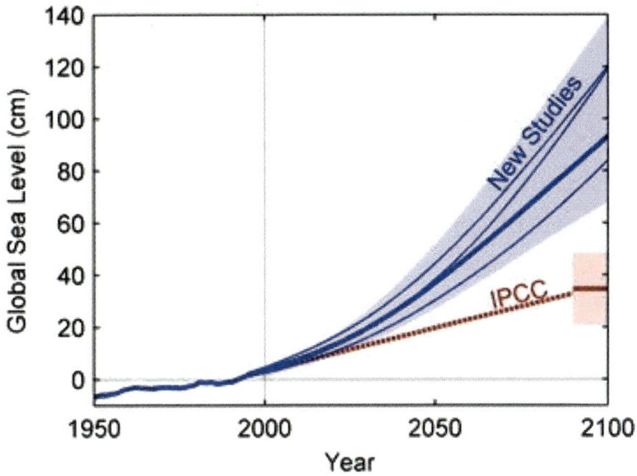

Fig. 2.10. Sea Level Rise Projections for the 21st Century.

Note: The pink box indicates the "most likely" scenario estimated by IPCC,[66] with a best estimate of 37 cm. The blue lines indicate recent empirical projections from several recent studies.[67–70]

Source: Results are described in Jevrejeva, S, JC Moore and A Grinsted (2010). How will sea level respond to changes in natural and anthropogenic forcings by 2100? *Geophysical Research Letters*, 37, L07703. Image from http://www.glaciology.net/Home/PDFs/Announcements [accessed 13 April 2011].

It is worrisome to base projections upon simple statistical relations, devoid of physics, when there are nonlinear processes involved. Ice sheets are clearly more climatically sensitive than previously understood, but the dynamical response time of ice sheets remains long relative to the forcing: decades to millennia for some aspects of ice sheet adjustment to a step change in climate. There are also threshold processes at play, such as ice shelf breakup and the transition from melt-free to melt-affected portions of the ice sheets as temperatures cross 0°C. Rates of sea level rise are unlikely to be a simple function of temperature. In fact, this assumption is akin to modeling global average temperature as an empirical function of CO_2 concentrations. There is a correlation, but the climate system is not that simple, therefore, such a treatment neglects what is known about the physics of the system.

It is thus difficult to evaluate the best methodology to forecast future sea level rise.[71] The IPCC projections may provide a reasonable estimate of the "most likely" scenario, but they do not reflect the full extent of possible sea level rise. The lead authors of a recent report commissioned by the Arctic Council, the Arctic Monitoring and Assessment Project, conclude that 90–160 cm of sea level rise is likely by 2100, based on the empirical results cited above, the dramatic thinning of Alaskan glaciers in recent decades, and the potential for destabilization of the Greenland Ice Sheet.

Estimates will probably be less in the pending IPCC Fifth Assessment Report, since the extent and rate of sea level rise from ice sheet destabilization is mostly a matter of speculation; the Greenland Ice Sheet may be able to react that quickly to climate warming, but it is not clear that such a collapse is the most likely scenario. For the Greenland Ice Sheet to contribute 50 cm to global sea level by 2100, it will have to experience a roughly 10-fold increase in its recent rate of mass loss. Such scenarios have been explored by examining the discharge from all of the major marine-based outlet glaciers of the Greenland Ice Sheet.[72] By taking the cross-sectional area of these outlet glaciers and assuming a constant ice thickness at these "flux gates", a doubling of ice speed would contribute 9 cm to global sea level by 2100.[73] Adding this to the sea level rise associated with surface mass balance losses gives an upper bound of about 20 cm for Greenland's contributions to sea level rise this century.[74]

While the estimates will remain controversial for some time, there is a clear urgency to better understand and quantify the risk of decadal-to century-scale sea level rise. There are concerted efforts to improve the representation of ice dynamics and ice-climate interactions in the next generation of models, but until this is realized, there will be considerable uncertainty in sea level projections. One thing that is clear is that warmer temperatures will cause global sea level rise at ever-increasing rates in the coming decades. Contributions from glacier and ice sheet melt are likely to eclipse those due to thermal expansion.

2.6. REGIONAL SEA LEVEL

As noted above, global eustatic sea level rise is not being experienced equally throughout the world. Several factors influence local and regional sea levels. Where ice sheets weighed down the landscape during the last glaciation, which ended ca. 10,000 years ago, the land is still recovering: rising slowly through the process of isostatic rebound. This partially counters warming-driven sea level rise in some regions. Land is springing up through the same process as a consequence of recent glacier and ice sheet retreat in places like Patagonia, Alaska, Iceland, and Greenland.

More subtle than this are patterns of regional sea level rise associated with the Earth's gravitational field. As ice is lost from areas such as Alaska and Greenland, there is less regional surface mass exerting a gravitational attraction to water in the oceans. Less water is drawn to these areas, and instead water in the oceans "piles up" in regions far from the polar icefields: the tropics. Hence, as ice mass is lost in both the Arctic and Antarctica, there is relatively less sea level rise in the polar regions and greater sea level rise in the tropics. The effect can be significant, e.g., more than 1.2 mm of sea level rise in the tropical Pacific for every mm of eustatic sea level equivalent derived from Greenland or West Antarctica.[75,76]

There are related geophysical effects on the global distribution of water in the oceans associated with the Earth's rotation; changes in the Earth's center of mass induce changes in the rate of rotation, decreasing or increasing Earth's "equatorial bulge". These effects are probably minor for small amounts of sea level rise.[77]

More significant are regional changes associated with thermostatic (density) effects, which result from ocean warming (thermal expansion) and changes in salinity. The world's oceans are warming at different rates and to different depths, affecting the extent of regional sea level rise.[78] Changes in ocean circulation also result from changes in air pressure and winds, and can influence the distribution of water in the global oceans. El Ni˜no-Southern Oscillation (ENSO) cycles offer a good example. On inter-annual time scales, ENSO variability shifts the strength and location of the West Pacific Warm Pool. During La Ni˜na, the easterly trade winds are strengthened and the warm pool is unusually well-developed, giving relatively high sea levels in the Western tropical Pacific; this is the time when countries like Indonesia, Malaysia, and Singapore are most vulnerable to the impacts of rising seas, as ENSO variability is superimposed on the background trend of higher seas due to long-term warming and glacial runoff.

Overall, these regional effects combine to make sea level rise a greater issue in the tropical regions than at high latitudes.[79] For a global eustatic sea level rise of 1 meter, Singapore and neighboring countries in the tropical Pacific and the Indian Ocean would likely experience sea level rise of 1.2–1.5 meters, depending on the extent of regional thermostatic change. Combined with the extensive land mass and infrastructure at low elevations in Singapore, the impact of sea level rise is likely to be extensive.[80]

2.7. CONCLUSION

The Arctic has often been referred to as the world's canary with respect to global warming, and this analogy is apt in many ways. Arctic amplification of global temperature variability occurs due to a variety of positive feedback processes in the Arctic. As a result, climate warming in the past few decades has been strongly manifested in the Arctic region; some of the climate changes that are projected to arrive in much of the world at some point this century are already underway in the Arctic. This is causing melting and loss of sea ice, permafrost, alpine glaciers, and the Greenland Ice Sheet.

Sea ice decline is most notable in the late-summer (August and September). The area of open water in the Arctic at this time of year is beginning to look enticing for commercial activity and transportation routes. Indeed, the Arctic appears to be on a trajectory towards seasonally ice-free conditions within a few decades. Thick multiyear ice that piles up against the northern coast of Greenland and the Canadian Arctic Archipelago is likely to persist, as will winter sea ice throughout the Arctic basin, but these may not interfere with a short and productive summer season where Arctic shipping routes are safe and accessible.

It is not yet clear how soon these routes will become viable, nor is it clear how reliable they will be; sea ice has a rapid response time with respect to climate, so it can re-establish itself quickly over a few cold years. Where there are ice floes present in the "shoulder season", storms and wind shifts can also bring this ice into the shipping lanes where it poses a hazard to ship traffic. There is nevertheless some multiyear persistence of sea ice conditions, associated in part with warming of the ocean and thinning of multiyear ice when there is more open water. This creates the possibility for good medium-range forecasting skill for ice conditions.

Large areas of the continental shelf are also seasonally open to resource exploration and extraction, and the length of the summer/fall season that permits economic activity is increasing. This reduces costs and expands the possibilities for drilling and mining operations in the Arctic. Given this opening window, increasing global demand, the need for alternative supplies, and improving technological capabilities, Arctic resource development is certain to increase in the coming decades. Increased infrastructure such as deepwater ports and railways will make this more attractive. However, large logistical challenges will remain for year-round activity, and the same caveats for shipping — inter-annual variability and shifting pack ice conditions — will ensure a challenging offshore work environment for the next few decades. Improved ice observations and short-to medium-range forecasts will help to alleviate costs and risks.

As sea ice declines and the ocean warms, increased storm and wave activity is also expected. This will create hazards for offshore and coastal infrastructure, and development plans need to accommodate this. These oceanographic changes are already causing increased coastal erosion and negative impacts on outlet glaciers of Arctic ice fields that are in contact with the sea. For marine outlet glaciers, the consequent thinning is causing ice shelf breakup and accelerated flow, leading to increased ice discharge. This thinning and acceleration is propagating inland and is likely to continue in some parts of the ice sheet. As a result of increased glacial melt and accelerated flow, rates of global sea level rise have increased. The projections for 21st century sea level rise from the IPCC are broadly considered to be too conservative, with recent estimates from several different studies of 80–160 cm by 2100. These are about three times greater than the IPCC's "best estimate" of 37 cm.

The truth may be somewhere in the middle of the IPCC projections and the recent, empirically-based estimates, as none of the estimates to date have a firm physical foundation. IPCC values are based on ice sheet models that are known to be too conservative because there is not yet a full understanding of much of the essential physics, such as the equations and environmental controls of iceberg calving and fast basal flow. Current models lack the capability to model ice sheet-ocean interactions and fast-flow dynamics. Ice-ocean coupling appears to be driving recent changes in both Greenland and West Antarctica, so this is a priority to understand and include in model projections, if models are to have reasonable predictive skills. On the other side, the higher estimates of sea level rise are based on empirical relationships to global temperature, neglecting the physics of the system and the potential for long timescales of ice sheet response for at least parts of the ice sheet system. Major destabilization of Greenland and/or Antarctica would need to occur for a sea level rise in excess of one meter by 2100; this would require a tripling of the current rate of global sea level rise ($3.2\,\mathrm{mm\,yr^{-1}}$), beginning immediately and sustained through the century. This is not impossible but it is easier to imagine a gradual increase in the rate of sea level rise throughout the century, perhaps reaching rates of $10\,\mathrm{mm\,yr^{-1}}$ after several decades.

It is possible that the values at the upper end of the IPCC projections (ca. 50–60 cm) represent a good estimate of the "most likely" scenario this century, while the recent estimates of 80–160 cm may better represent the upper tail of the probability distribution: thus a physically-possible but low-probability scenario. Impacts associated with these upper bounds are obviously quite grave, however, so they might offer reasonable, conservative values for some policy decisions. For many decisions, the "most likely" estimate is not as important as the "worst case" scenario at the low-probability, upper end of the probability distribution. Climate-change policy, emergency response plans for tropical cyclones, and coastal engineering projects all need to consider both the most likely outcomes and the extreme (but physically possible) scenarios with respect to sea level rise. This is particularly true for low-lying tropical nations, where the impacts and extent of sea level rise will be most severe.

BIBLIOGRAPHY

Alley, RB, PU Clark, P Huybrechts and I Joughin (2005). Ice-sheet and sea-level changes. *Science*, 310, 456–460.

AMAP (2009). *The Greenland Ice Sheet in a Changing Climate: Snow, Water, Ice and Permafrost in the Arctic (SWIPA)*. D Dahl-Jensen *et al.*, Arctic Monitoring and Assessment Program, Oslo, Norway, pp. 115.

Box, JE, DH Bromwich, BA Veenhuis, L-S Bai, JC Stroeve, JC Rogers, K Steffen, T Haran and S-H Wang (2006). Greenland ice sheet surface mass balance variability (1988–2004) from calibrated polar MM5 output. *Journal of Climate*, 19(12), 2783–2800.

Church, JA and NJ White (2011). Sea-level rise from the late 19th to the early 21st century. *Surveys in Geophysics*, 32, 585–602, doi:10.1007/s10712-011-9119-1.

CIA Factbook. https://www.cia.gov/library/publications/the-world-factbook/geos/ho.html [accessed 8 January 2011]. Here I consider urban centers with populations exceeding 2 million that are coastal or connected to the ocean via a major waterway, e.g., the Thames River.

Cuffey, KM and SJ Marshall (2000). Sea level rise from Greenland ice sheet retreat in the last interglacial period. *Nature*, 404, 591–594.

Das, SB, I Joughin, MD Ben, IM Howat, MA King, D Lizarralde and MP Bhatia (2008). Fracture propagation to the base of the Greenland Ice Sheet during supraglacial lake drainage. *Science*, 320 (5877), 778–781.

Dasgupta, S, B Laplante, C Meisner, D Wheeler and J Yan (2007). The impact of sea-level rise on developing countries: A comparative analysis. World Bank working paper.

Ettema, J, MR van den Broeke, E van Meijgaard, WJ van de Berg, JL Bamber, JE Box and RC Bales (2009). Higher surface mass balance of the Greenland ice sheet revealed by high-resolution climate modeling. *Geophysical Research Letters*, 36, L12501, doi:10.1029/2009GL038110.

Fetterer, F, K Knowles, W Meier and M Savoie (2002, updated 2011). *Sea Ice Index*. Boulder, CO: National Snow and Ice Data Center. Digital media. http://nsidc.org/data/seaice index/index.html [accessed 10 July 2011].

Giles, KA, SW Laxon and AL Ridout (2008). Circumpolar thinning of Arctic sea ice following the 2007 record ice extent minimum. *Geophysical Research Letters*, 35, L22502.

Gregory, JM and P Huybrechts (2006). Ice-sheet contributions to future sea-level change. *Philosophical Transactions of the Royal Society, Series A*, 364, 1709–1731.

Grinsted, A, JC Moore and S Jevrejeva (2009). Reconstructing sea level from paleo and projected temperatures 200 to 2100 AD. *Climate Dynamics*, 34, 461–472, doi:10.1007/s00382-008-0507-2.

Holland, DM, RH Thomas, B De Young, MH Ribergaard and B Lyberth (2008). Acceleration of Jakobshavn Isbrae triggered by warm subsurface ocean waters. *Nature Geoscience* , 1, 659–664.

Holland, MM and CM Bitz (2003). Polar amplification of climate change in coupled models. *Climate Dynamics*, 21, 221–232.

Holland, MM, CM Bitz and B Tremblay (2006). Future abrupt reductions in the summer Arctic sea ice. *Geophysical Research Letters*, 33, L23503, doi:10.1029/2006GL028024.

Holland, MM, MC Serreze and J Stroeve (2010). The sea ice mass budget of the Arctic and its future change as simulated by coupled climate models. *Climate Dynamics*, 34(2), 185–200, doi:10.1007/s00382-008-0493-4.

Hunter, JR (2010). Estimating sea-level extremes under conditions of uncertain sea-level rise. *Climatic Change*, 99, 331–350, doi:10.1007/s10584009-9671-6.

Huybrechts, P and J de Wolde (1999). The dynamic response of the Greenland and Antarctic ice sheets to multiple-century climatic warming. *Journal of Climate*, 12, 2169–2188.

Huybrechts, P, A Letr'eguilly and N Reeh (1991). The Greenland ice sheet and greenhouse warming. *Palaeogeography, Palaeoclimatology, Palaeoecology*, 89, 399–412.

Huybrechts, P, J Gregory, I Janssens and M Wild (2004). Modelling Antarctic and Greenland ice volume changes during the 20th and 21st centuries forced by GCM time slice integrations. *Global and Planetary Change*, 42, 83–105.

IPCC (2007). Summary for Policymakers. In *Climate Change 2007: The Physical Science Basis. Contribution of Working Group I to the Fourth Assessment Report of the Intergovernmental Panel on Climate Change*, S Solomon, D Qin, M Manning, Z Chen, M Marquis, KB Averyt, M Tignor and HL Miller (eds.). Cambridge, United Kingdom and New York, NY, USA: Cambridge University Press.

Jevrejeva, S, JC Moore and A Grinsted (2010). How will sea level respond to changes in natural and anthropogenic forcings by 2100. *Geophysical Research Letters*, 37(L07703), doi:10.1029/2010GL042947.

Joughin, I, SB Das, MA King, BE Smith, IM Howat and T Moon (2008). Seasonal speedup along the Western flank of the Greenland ice sheet. *Science*, 320(5877), 781–783.

Kaser, G, JG Cogley, MB Dyurgerov, MF Meier and A Ohmura (2006). Mass balance of glaciers and ice caps: Consensus estimates for 1961–2004. *Geophysical Research Letters*, 33(19), L19501.

Kwok, R and DA Rothrock (2009). Decline in Arctic sea ice thickness from submarine and ICES at records: 1958–2008. *Geophysical Research Letters*, 36, L15501.

Lemke, P, J Ren, RB Alley, I Allison, J Carrasco, G Flato, Y Fujii, G Kaser, P Mote, RH Thomas and T Zhang (2007). Observations: Changes in snow, ice and frozen ground. In *Climate Change 2007: The Physical Science Basis. Contribution of Working Group I to the Fourth Assessment Report of the Intergovernmental Panel on Climate Change*, S Solomon, D Qin, M Manning, Z Chen, M Marquis, KB Averyt, M Tignor and HL Miller (eds.). Cambridge, United Kingdom and New York, NY, USA: Cambridge University Press.

Marshall, SJ (2011). *The Cryosphere*, p. 288. Princeton, NJ: Princeton. *Primers in Climate* , Princeton University Press.

Milne, GA and JX Mitrovica (2008). Searching for eustasy in deglacial sea-level histories. *Quaternary Science Reviews*, 27, 2292–2302.

Milne, GA, WR Gehrels, CW Hughes and ME Tamisiea (2009). Identifying the causes of sea-level change. *Nature Geoscience*, 2, 471–478.

Mitrovica, JX, N Gomez and PU Clark (2009). The sea-level fingerprint of West Antarctic collapse. *Science*, 323(5915), 753.

Mote, TL (2007). Greenland surface melt trends 1973–2007: Evidence of a large increase in 2007. *Geophysical Research Letters*, 34, L22507, doi:10.1029/2007GL031976.

Ng, W-S and R Mendelsohn (2005). The impact of sea level rise on Singapore. *Environment and Development Economics*, 10, 201–215, doi:10.1017/S1355770X04001706.

Nicholls, RJ and A Cazenave (2010). Sea-level rise and its impact on coastal zones. *Science*, 328(5985), 1517–1520, doi:10.1126/science.1185782.

Nick, FM, A Vieli, IM Howat and I Joughin (2009). Large-scale changes in Greenland outlet glacier dynamics triggered at the terminus. *Nature Geoscience*, 2, 110–114.

Otto-Bliesner, B, SJ Marshall, JT Overpeck, GH Miller, G Hu and CAPE Last Interglacial Project members (2006). Simulating Arctic climate warmth and ice sheet retreat in the last interglaciation. *Science*, 311, 1751–1753.

Payne, AJ, A Vieli, A Shepherd, DJ Wingham and E Rignot (2004). Recent dramatic thinning of largest West-Antarctic ice stream triggered by oceans. *Geophysical Research Letters*, 31, L23401.

Pfeffer, WT, JT Harper and S O'Neel (2008). Kinematic constraints on glacier contributions to 21st century sea-level rise. *Science*, 321, 1340–1343.

Pritchard, HD, RJ Arthern, DG Vaughan and LA Edwards (2009). Extensive dynamic thinning on the margins of the Greenland and Antarctic ice sheets. *Nature*, 461, 971–975.

Rahmstorf, S (2007). A semi-empirical approach to projecting future sea-level rise. *Science*, 315(5810), 368–370.

Rahmstorf, S, A Cazenave, JA Church, JE Hansen, RF Keeling, DE Parker and RCJ Somerville (2007). Recent climate observations compared to predictions. *Science*, 316, 709.

Reeh, N, C Mayer, H Miller, HH Thomsen and A Weidick (1999). Present and past climate control on fjord glaciations in Greenland: Implications for IRD-deposition in the sea. *Geophysical Research Letters*, 26, 1039–1042.

Rignot, EJ and RH Thomas (2002). Mass balance of polar ice sheets. *Science*, 297, 1502–1506.

Rignot, E and P Kanagaratna (2006). Changes in the velocity structure of the Greenland ice sheet. *Science*, 311, 986–990.

Stroeve, J, MM Holland, W Meier, T Scambos and M Serreze (2007). Arctic sea ice decline: Faster than forecast. *Geophysical Research Letters*, 34, L09501, doi:10.1029/2007GL029703.

SWIPA (2011). *Snow, Water, Ice and Permafrost in the Arctic (SWIPA) Executive Summary.* Arctic Monitoring and Assessment Program, Oslo, Norway.

Tedesco, M, X Fettweis, MR van den Broeke, RSW van de Wal, CJPP Smeets, WJ van de Berg, MC Serreze and JE Box (2011). The role of albedo and accumulation in the 2010 melting record in Greenland. *Environmental Research Letters*, 6(014005), doi: 10.1088/1748-9326/6/1/014005.

Van den Broeke, M, J Bamber, J Ettema, E Rignot, E Schrama, WJ van de Berg, E van Meijgaard, I Velicogna and B Wouters (2009). Partitioning recent Greenland mass loss. *Science*, 326(5955), 984–986.

Van der Veen, CJ (2002). Polar ice sheets and global sea level: How well can we predict the future. *Global and Planetary Change*, 32(2–3), 165–194.

Vaughan, DG and R Arthern (2007). Why is it hard to predict the future of ice sheets. *Science*, 315(5818), 1503–1504.

Velicogna, I (2009). Increasing rates of ice mass loss from the Greenland and Antarctic ice sheets revealed by GRACE. *Geophysical Research Letters*, 36, L19503, doi:10.1029/2009GL040222.

Vermeer, M and S Rahmstorf (2009). Global sea level linked to global temperature. *Proceedings of the National Academy of Science of the USA*, 106, 21,527–21,532.

Zhang, X and JE Walsh (2006). Toward a seasonally ice-covered Arctic ocean: Scenarios from the IPCC AR4 model simulations. *Journal of Climate*, 19, 1730–1747.

Zwally, HJ, W Abdalati, T Herring, K Larson, J Saba and K Steffen (2002). Surface melt-induced acceleration of Greenland ice-sheet flow. *Science*, 297, 218–222.

ENDNOTES

1 Fetterer, F, K Knowles, W Meier and M Savoie (2002, updated 2011). *Sea Ice Index.* Boulder, CO: National Snow and Ice Data Center, Digital media. http://nsidc.org/data/seaice index/index.html [accessed 10 July 2011].

2 Stroeve, J, MM Holland, W Meier, T Scambos and M Serreze (2007). Arctic sea ice decline: Faster than forecast. *Geophysical Research Letters*, 34, L09501, doi:10.1029/2007GL029703.

3 Giles, KA, SW Laxon and AL Ridout (2008). Circumpolar thinning of Arctic sea ice following the 2007 record ice extent minimum. *Geophysical Research Letters*, 35, L22502.

4 Kwok, R and DA Rothrock (2009). Decline in Arctic sea ice thickness from submarine and ICES at records: 1958–2008. *Geophysical Research Letters*, 36, L15501.

5 Lemke, P, J Ren, RB Alley, I Allison, J Carrasco, G Flato, Y Fujii, G Kaser, P Mote, RH Thomas and T Zhang (2007). Observations: Changes in snow, ice and frozen ground. In *Climate Change 2007: The Physical Science Basis. Contribution of Working Group I to the Fourth Assessment Report of the Intergovernmental Panel on Climate Change*, S Solomon, D Qin, M Manning, Z Chen, M Marquis, KB Averyt, M Tignor and HL Miller (eds.). Cambridge, United Kingdom and New York, NY, USA: Cambridge University Press.

6 SWIPA (2011). *Snow, Water, Ice and Permafrost in the Arctic (SWIPA) Executive Summary.* Arctic Monitoring and Assessment Programme, Oslo, Norway

7 Marshall, SJ (2012). *The Cryosphere*, p. 288. Princeton, NJ: Princeton Primers in Climate, Princeton University Press.

8 Kaser, G, JG Cogley, MB Dyurgerov, MF Meier and A Ohmura (2006). Mass balance of glaciers and ice caps: Consensus estimates for 1961–2004. *Geophysical Research Letters*, 33(19), L19501.

9 Mote, TL (2007). Greenland surface melt trends 1973–2007: Evidence of a large increase in 2007. *Geophysical Research Letters*, 34, L22507, doi:10.1029/2007GL031976.

10 Tedesco, M, X Fettweis, MR van den Broeke, RSW van de Wal, CJPP Smeets, WJ van de Berg, MC Serreze and JE Box (2011). The role of albedo and accumulation in the 2010 melting record in Greenland. *Environmental Research Letters*, 6(014005), doi: 10.1088/1748-9326/6/1/014005.

11 Rignot, E and P Kanagaratnam (2006). Changes in the velocity structure of the Greenland ice sheet. *Science*, 311, 986–990.

12 Pritchard, HD, RJ Arthern, DG Vaughan and LA Edwards (2009). Extensive dynamic thinning on the margins of the Greenland and Antarctic ice sheets. *Nature*, 461, 971–975.

13 Dasgupta, S, B Laplante, C Meisner, D Wheeler and J Yan (2007). The impact of sea-level rise on developing countries: A comparative analysis. World Bank working paper.

14 CIA Factbook. Here I consider urban centers with populations exceeding 2 million that are coastal or connected to the ocean via a major waterway, e.g., the Thames River. https://www.cia.gov/library/publications/the-world-factbook/geos/ho.html [accessed 8 January 2011].

15 Holland, MM and CM Bitz (2003). Polar amplification of climate change in coupled models. *Climate Dynamics*, 21, 221–232.

16 Fetterer, F, K Knowles, W Meier and M Savoie (2002, updated 2011). *Sea Ice Index*. Boulder, CO: National Snow and Ice Data Center, Digital media. http://nsidc.org/data/seaice index/index.html [accessed 10 July 2011], *op.cit.*

17 Zhang, X and JE Walsh (2006). Toward a seasonally ice-covered Arctic ocean: Scenarios from the IPCC AR4 model simulations. *Journal of Climate*, 19, 1730–1747.

18 Holland, MM, CM Bitz and B Tremblay (2006). Future abrupt reductions in the summer Arctic sea ice. *Geophysical Research Letters*, 33, L23503, doi:10.1029/2006GL028024.

19 Holland, MM, MC Serreze and J Stroeve (2010). The sea ice mass budget of the Arctic and its future change as simulated by coupled climate models. Climate Dynamics, 34(2), doi:10.1007/s00382-008-0493-4.

20 Stroeve, J, MM Holland, W Meier, T Scambos and M Serreze (2007). Arctic sea ice decline: Faster than forecast. *Geophysical Research Letters*, 34, L09501, doi:10.1029/2007GL029703, 2007, *op.cit.*

21 Giles, KA, SW Laxon and AL Ridout (2008). Circumpolar thinning of Arctic sea ice following the 2007 record ice extent minimum. *Geophysical Research Letters*, 35, L22502, *op.cit.*

22 Kwok, R and DA Rothrock (2009). Decline in Arctic sea ice thickness from submarine and ICES at records: 1958–2008. *Geophysical Research Letters*, 36, L15501, *op.cit.*

23 Tedesco, M, X Fettweis, MR van den Broeke, RSW van de Wal, CJPP Smeets, WJ van de Berg, MC Serreze and JE Box (2011). The role of albedo and accumulation in the 2010 melting record in Greenland. *Environmental Research Letters*, 6(014005), doi: 10.1088/1748-9326/6/1/014005, *op. cit.*

24 Box, JE, DH Bromwich, BA Veenhuis, L-S Bai, JC Stroeve, JC Rogers, K Steffen, T Haran, S-H Wang (2006). Greenland ice sheet surface mass balance variability (1988–2004) from calibrated polar MM5 output. *Journal of Climate*, 19(12), 2783–2800.

25 AMAP (2009). *The Greenland Ice Sheet in a Changing Climate: Snow, Water, Ice and Permafrost in the Arctic* (*SWIPA*), D Dahl-Jensen

et al. (eds.), 115 p. Oslo, Norway: Arctic Monitoring and Assessment Program.

26 Ettema, J, MR van den Broeke, E van Meijgaard, WJ van de Berg, JL Bamber, JE Box and RC Bales (2009). Higher surface mass balance of the Greenland ice sheet revealed by high-resolution climate modeling. *Geophysical Research Letters*, 36, L12501, doi:10.1029/2009GL038110.

27 Box, JE, DH Bromwich, BA Veenhuis, L-S Bai, JC Stroeve, JC Rogers, K Steffen, T Haran, S-H Wang (2006). Greenland ice sheet surface mass balance variability (1988–2004) from calibrated polar MM5 output. *Journal of Climate*, 19(12), 2783–2800, *op. cit.*

28 AMAP (2009). *The Greenland Ice Sheet in a Changing Climate: Snow, Water, Ice and Permafrost in the Arctic (SWIPA)*. D Dahl-Jensen *et al.*, 115 p. Oslo, Norway: Arctic Monitoring and Assessment Program, *op. cit.*

29 Rignot, E and P Kanagaratnam (2006). Changes in the velocity structure of the Greenland ice sheet. *Science*, 311, 986–990, *op. cit.*

30 Reeh, N, C Mayer, H Miller, HH Thomsen and A Weidick (1999). Present and past climate control on fjord glaciations in Greenland: Implications for IRD-deposition in the sea. *Geophysical Research Letters*, 26, 1039–1042.

31 Velicogna, I (2009). Increasing rates of ice mass loss from the Greenland and Antarctic ice sheets revealed by GRACE. *Geophysical Research Letters*, 36, L19503, doi:10.1029/2009GL040222.

32 van den Broeke, M, J Bamber, J Ettema, E Rignot, E Schrama, WJ van de Berg, E van Meijgaard, I Velicogna and B Wouters (2009). Partitioning recent Greenland mass loss. *Science*, 326(5955), 984–986.

33 Holland, DM, RH Thomas, B De Young, MH Ribergaard and B Lyberth (2008). Acceleration of Jakobshavn Isbrae triggered by warm subsurface ocean waters. *Nature Geoscience*, 1, 659–664.

34 *Ibid.*

35 Nick, FM, A Vieli, IM Howat and I Joughin (2009). Large-scale changes in Greenland outlet glacier dynamics triggered at the terminus. *Nature Geoscience*, 2, 110–114.

36 Mote, TL (2007). Greenland surface melt trends 1973–2007: Evidence of a large increase in 2007. *Geophysical Research Letters*, 34, L22507, doi:10.1029/2007GL031976, *op. cit.*

37 Tedesco, M, X Fettweis, MR van den Broeke, RSW van de Wal, CJPP Smeets, WJ van de Berg, MC Serreze and JE Box (2011). The role of albedo and accumulation in the 2010 melting record in Greenland. *Environmental Research Letters*, 6(014005), doi: 10.1088/1748-9326/6/1/014005, *op. cit.*

38 Zwally, HJ, W Abdalati, T Herring, K Larson, J Saba and K Steffen (2002). Surface melt-induced acceleration of Greenland ice-sheet flow. *Science*, 297, 218–222.

39 Das, SB, I Joughin, MD Ben, IM Howat, MA King, D Lizarralde and MP Bhatia (2008). Fracture propagation to the base of the Greenland ice sheet during supraglacial lake drainage. *Science*, 320(5877), 778–781.

40 Joughin, I, SB Das, MA King, BE Smith, IM Howat and T Moon (2008). Seasonal speedup along the Western flank of the Greenland ice sheet. *Science*, 320(5877), 781–783.

41 Cuffey, KM and SJ Marshall (2000). Sea level rise from Greenland ice sheet retreat in the last interglacial period. *Nature*, 404, 591–594.

42 Otto-Bliesner, B, SJ Marshall, JT Overpeck, GH Miller, G Hu and CAPE Last Interglacial Project members (2006). Simulating Arctic climate warmth and ice sheet retreat in the last interglaciation. *Science*, 311, 1751–1753.

43 *Ibid.*

44 *Ibid.*

45 Huybrechts, P, A Letr'eguilly and N Reeh (1991). The Greenland ice sheet and greenhouse warming. *Palaeogeography, Palaeoclimatology, Palaeoecology*, 89, 399–412.

46 Alley, RB, PU Clark, P Huybrechts and I Joughin (2005). Ice-sheet and sea-level changes. *Science*, 310, 456–460.

47 Gregory, JM and P Huybrechts (2006). Ice-sheet contributions to future sea-level change. *Philosophical Transactions of the Royal Society, Series A*, 364, 1709–1731.

48 *Ibid.*

49 IPCC (2007). Summary for policymakers. In *Climate Change 2007: The Physical Science Basis. Contribution of Working Group I to the Fourth Assessment Report of the Intergovernmental Panel on Climate Change*, S Solomon, D Qin, M Manning, Z Chen, M Marquis, KB Averyt, M Tignor and HL Miller (eds.). Cambridge, United Kingdom and New York, NY, USA: Cambridge University Press.

50 Church, JA and NJ White (2011). Sea-level rise from the late 19th to the early 21st century. *Surveys in Geophysics*, doi:10.1007/s10712-0119119-1.

51 Rignot, EJ and RH Thomas (2002). Mass balance of polar ice sheets. *Science*, 297, 1502–1506.

52 Payne, AJ, A Vieli, A Shepherd, DJ Wingham and E Rignot (2004). Recent dramatic thinning of largest West-Antarctic ice stream triggered by oceans. *Geophysical Research Letters*, 31, L23401.

53 Church, JA and NJ White (2011). Sea-level rise from the late 19th to the early 21st century. *Surveys in Geophysics*, doi:10.1007/s10712-0119119-1.

54 IPCC (2007). Summary for policymakers. In *Climate Change 2007: The Physical Science Basis. Contribution of Working Group I to the Fourth Assessment Report of the Intergovernmental Panel on Climate Change*, S Solomon, D Qin, M Manning, Z Chen, M Marquis, KB Averyt, M Tignor and HL Miller (eds.). Cambridge, United Kingdom and New York, NY, USA: Cambridge University Press, *op. cit* .

55 Van der Veen, CJ (2007). Polar ice sheets and global sea level: How well can we predict the future. *Global and Planetary Change*, 32(2–3), 165–194.

56 Vaughan, DG and R Arthern (2007). Why is it hard to predict the future of ice sheets. *Science*, 315(5818), 1503–1504.

57 Rahmstorf, S (2007). A semi-empirical approach to projecting future sea-level rise. *Science*, 315(5810), 368–370.

58 Rahmstorf, S, A Cazenave, JA Church, JE Hansen, RF Keeling, DE Parker and RCJ Somerville (2007). Recent climate observations compared to predictions. *Science*, 316, 709.

59 Grinsted, A, JC Moore and S Jevrejeva (2009). Reconstructing sea level from paleo and projected temperatures 200 to 2100 AD. *Climate Dynamics*, doi:10.1007/s00382-008-0507-2.

60 Huybrechts, P and J de Wolde (1999). The dynamic response of the Greenland and Antarctic ice sheets to multiple-century climatic warming. *Journal of Climate*, 12, 2169–2188.

61 Huybrechts, P, J Gregory, I Janssens and M Wild (2004). Modelling Antarctic and Greenland ice volume changes during the 20th and 21st centuries forced by GCM time slice integrations. *Global and Planetary Change*, 42, 83–105.

62 Rahmstorf, S (2007). A semi-empirical approach to projecting future sea-level rise. *Science*, 315(5810), 368–370, *op. cit.*

63 Grinsted, A, JC Moore and S Jevrejeva (2009). Reconstructing sea level from paleo and projected temperatures 200 to 2100 AD. *Climate Dynamics*, doi:10.1007/s00382-008-0507-2, *op. cit.*

64 Vermeer, M and S Rahmstorf (2009). Global sea level linked to global temperature. *Proceedings of the National Academy of Science of the USA*, 106, 21, 527–21, 532.

65 Jevrejeva, S, JC Moore and A Grinsted (2010). How will sea level respond to changes in natural and anthropogenic forcings by 2100. *Geophysical Research Letters*, 37(L07703), doi:10.1029/2010GL042947.

66 IPCC (2007). Summary for policymakers. In *Climate Change 2007: The Physical Science Basis. Contribution of Working Group I to the Fourth Assessment Report of the Intergovernmental Panel on Climate Change*, S Solomon, D Qin, M Manning, Z Chen, M Marquis, KB Averyt,

M Tignor and HL Miller (eds.). Cambridge, United Kingdom and New York, NY, USA: Cambridge University Press, *op. cit.*

67 Rahmstorf, S (2007). A semi-empirical approach to projecting future sea-level rise. *Science*, 315(5810), 368–370, *op. cit.*

68 Grinsted, A, JC Moore and S Jevrejeva (2009). Reconstructing sea level from paleo and projected temperatures 200 to 2100 AD. *Climate Dynamics*, doi:10.1007/s00382-008-0507-2, *op. cit.*

69 Vermeer, M and S Rahmstorf (2009). Global sea level linked to global temperature. *Proceedings of the National Academy of Science of the USA*, 106, 21, 527–21, 532, *op. cit.*

70 Jevrejeva, S, JC Moore and A Grinsted (2010). How will sea level respond to changes in natural and anthropogenic forcings by 2100. *Geophysical Research Letters*, 37(L07703), doi:10.1029/2010GL042947, *op. cit.*

71 Hunter, JR (2010). Estimating sea-level extremes under conditions of uncertain sea-level rise. *Climatic Change*, 99, 331–350, doi:10.1007/s10584-009-9671-6.

72 Pfeffer, WT, JT Harper and S O'Neel (2008). Kinematic constraints on glacier contributions to 21st century sea-level rise. *Science*, 321, 1340–1343.

73 *Ibid.*

74 AMAP (2009). *The Greenland Ice Sheet in a Changing Climate: Snow, Water, Ice and Permafrost in the Arctic* (*SWIPA*). D Dahl-Jensen *et al.*, 115 p. Oslo, Norway: Arctic Monitoring and Assessment Programme, *op. cit.*

75 Milne, GA, WR Gehrels, CW Hughes and ME Tamisiea (2009). Identifying the causes of sea-level change. *Nature Geoscience*, 2, 471–478.

76 Mitrovica, JX, N Gomez and PU Clark (2009). The sea-level fingerprint of West Antarctic collapse. *Science*, 323(5915), 753.

77 Milne, GA and JX Mitrovica (2008). Searching for eustasy in deglacial sea-level histories. *Quaternary Science Reviews*, 27, 2292–2302.

78 Church, JA and NJ White (2011). Sea-level rise from the late 19th to the early 21st century. *Surveys in Geophysics*, doi:10.1007/s10712-0119119-1, *op. cit.*

79 Nicholls, RJ and A Cazenave (2010). Sea-level rise and its impact on coastal zones. *Science*, 328(5985), 1517–1520, doi:10.1126/science.1185782.

80 Ng, W-S and R Mendelsohn (2005). The impact of sea level rise on Singapore. *Environment and Development Economics*, 10, 201–215, doi:10.1017/S1355770X04001706.

3

POSSIBLE CHANGES TO THE RUSSIAN ARCTIC ENVIRONMENT UNDER THE INFLUENCE OF NATURAL AND ANTHROPOGENIC FACTORS

Vyacheslav Mikhailovich Makeev

3.1. INTRODUCTION

The Russian Federation has particular national strategic interests in its Arctic zone relating to the mineral resources and the transport and defence potential for the country, and maintenance of the biosphere. The area of this zone is 9 million km^2 and its population is 2.5 million (2% of the Russian population), accounting for 40% of the Arctic population. It accounts for 12–15% of the gross domestic product (GDP) of the country and one fourth of Russia's exports. The scale of its economic activity considerably surpasses that of the other Arctic countries. Two-thirds of the whole Arctic wealth belongs to Russia. The estimated value of the Arctic's mineral resources is more than 30 trillion dollars, and two-thirds of this is in the form of energy resources.[1]

To a large extent, the exploration and export of mineral and energy resources, as well as the sustainable social and economic development of Russia's Arctic region depends on how efficiently the main arterial road of the Arctic — the Northern Sea Route (NSR) — will work. Economically-motivated functioning of this seaway is affected by many factors, the most important of which is the state of the ice cover during the navigation period. Warming of the climate in many regions of the Arctic over the last 30 years has led to the reduction of the Arctic ice and changes in other characteristics (thickness, age) of the ice cover in the Arctic Ocean (AO), which have greatly improved the conditions for and duration of navigation.

The question, "whether the navigation along the NSR in the near future will improve or worsen" does not yet have a clear cut answer. This uncertainty is caused by the various researchers' different understanding of the reasons underlying the climate change in global and regional climate systems.

This author links the current climate change with rhythmically repetitive solar-planetary processes, and therefore believes that in the near future a substantial improvement in the navigation conditions along the NSR will not occur. He in fact believes that the conditions will more likely become worse.

3.2. CHANGE OF CLIMATE

Although the data in the reports of the Intergovernmental Panel on Climate Change (IPCC) testifies the global character of the climate warming (2001–2007),[2] the experts behind those reports have repeatedly underlined that the warming has many regional features, which have to be investigated in order to understand the climate change better. Until now, the reasons for this warming are not absolutely clear; as a result, it is not known how long it will last.

In Russia, with its huge territory and natural variety, climate changes in different regions are specific enough. According to the Federal Hydrometeorology and Environmental Monitoring Service statistics, the average annual ground air temperature across Russia has grown by 1°C over the last 100 years. Between 1990 and 2000, the warming became considerably stronger and grew by 0.4°C (the strategic forecast, 2005).[3]

Another feature of the global warming at the end of the 20th century was its seasonal and regional differences. It affected winter and spring seasons and, to a smaller extent, autumn and summer.

Greater warming has been very apparent beyond the Ural Mountains compared to the European part. In the Russian Arctic, warming has affected the largest territory, but there are certain areas which remain untouched, or have even become cooler (i.e., the Western part of the Barents Sea, Kara Sea with Novaya Zemlya, the delta of the Lena River, and the Western part of Chukchi Peninsula). The same is observed in the Arctic zone of North America where there are areas experiencing warming (for example, Alaska), as well as areas with a steady cold snap (for example, the Northwestern Territory and Baffin Sea).

For prognostic purposes, it is very important to estimate the place and status of the modern climatic episode in the chain of non-uniformly scaled long-term climatic fluctuations in the Arctic region. For this purpose, we have studied the materials on changeability of air temperatures in the Arctic region during the Holocene (i.e., the geological period which started about 10,000 years ago after the last ice age), the last 1,000 years (based on paleogeographical and paleolimnogical researches)[4–6] and the last 70 years (done through instrumental observations). The analysis of these materials has shown a complicated structure of space and time variability of the Russian Arctic's climate (as well as in other areas) over the last 10,000–12,000 years as described below.

Although the cyclic character of temperature fluctuations does not differ everywhere, the sequence, orientation and amplitude of fluctuations, and the beginning and the end of cycle phases in various areas differ considerably. It is possible to identify three climatic provinces where the character of air temperature fluctuations covering several centuries (basically the summer season) is close to each other. They are: (a) the European climatic province, (b) the East Asian Arctic province, and (c) the West Asian Arctic province (Fig. 3.1).

The European Arctic province occupies the entire territory of the European North up to the Polar Urals, including the archipelagos Spitsbergen, Franz Josef Land, and Novaya Zemlya. It is characterized by the cyclic temperature fluctuations covering several centuries with warming phases prevailing over cold snap phases from 200 to

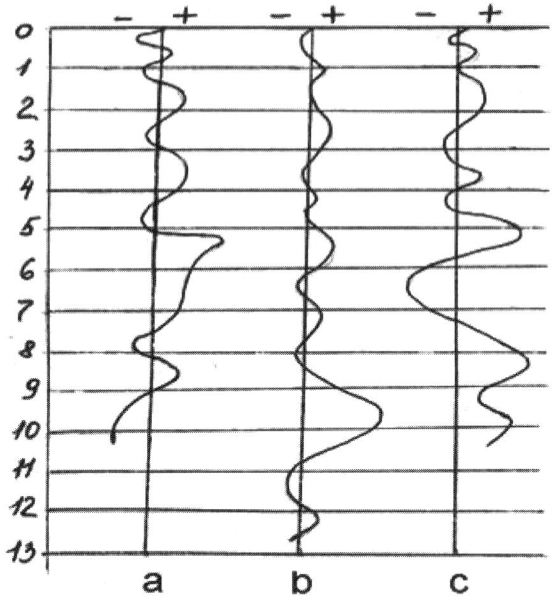

Fig. 3.1. Paleoclimate Curves Which Characterize Air Temperature Fluctuations Relative to the Contemporary Situation in Different Provinces of the Arctic.

Notes:
(a) The European Arctic climatic province,
(b) The East Asian Arctic province (Novosibirsk islands),
(c) The West Asian Arctic province (Taimyr).

Source: Mihailovich, MV (2011). The long-term changes of a thermal regime of the Arctic. *Izvestia of Russian Geographical Society*, 1, 28–35.

1,300 years and the air temperature amplitude of 4–6°C between the warmest and coldest period.

The air temperature changes of the Holocene period are characterized by gradual warming during the postglacial time up to the climatic optimum (about 5,500–6000 years ago) after which the fluctuations did not reach such amplitude and, as a whole, the climate has become colder compared to the average Holocene temperatures. Now this climatic province has passed a cold phase and is at the beginning of a warm phase. The East Asian Arctic province, covering the areas of the Northern Land Archipelago, the Novosibirsk

Islands, the Wrangel Islands, the most part of Chukchi Peninsula, and the Northern regions of East and Middle Siberia are characterized by longer phases (up to 2,000 years) of temperature fluctuations, with a greater amplitude of air temperatures between warm and cold phases (to 8–9°C), displaying the climatic optimum 10,000–8,800 years ago. Temperature curves help to distinguish earlier Holocene periods, i.e., less warm average Holocene and the coldest — top Holocene. Fluctuations in the summer temperatures for the average and late Holocene did not exceed 2–3°C. At present, the climatic province is at the beginning of a warm phase.

For the West Asian Arctic province, covering the territories of Yamal, Tazovsky and Gydansky Peninsulas, and a considerable part of the Taimyr Peninsula, in the super century, the occurrence of 1800–1900 summer rhythms with rather short exothermic phases is observed (200–400 years). In addition, longer humid phases at the end of which the climate cold snap often occurred are also observed. Generally, for this province, the warmer Holocene is characteristic, unlike the European Arctic regions where the lower Holocene, especially warm average Holocene (the climatic optimum is dated for an interval 6–5 thousand years ago), and colder top Holocene dominate. Now the province is at the beginning of a warm, and most likely, exothermic phase.

Generally, in the structure of temperature fluctuations in the Arctic, several time-cycles are distinguishable. Their duration ranges from 1,200 to 4,000 years. The beginning and end of warming and cold snaps in different areas could occur simultaneously and at varying times (up to their full asynchronicity). As a good example of asynchronicity, we can refer to the occurrence of a Holocene climatic optimum in the European and East Asian climatic provinces.

In the East Asian province, the Holocene optimum occurred 9,700–8,700 years ago and in the European province, only 6–5 thousand years ago. In both cases, the summer temperature increased, compared with the present day, by 4–5°C. This led to the essential re-organization of landscapes and the intensification of various processes weakly shown in the colder climate 300–400 km to the North, the activization of thermokarstic[7] and termoabrasion processes,[8] and melting of glaciers, fields, and seas. Among other examples of asynchronicity, there are intervals that occurred 1.5–2 thousand years ago and 2,500 years ago.

Synchronous climate changes in the Arctic region appeared in the following cycles: cold snap (10,500–11,000 years ago), warm snaps (9,700 thousand years ago, and 4,000 years ago and in the 9th–14th centuries) and a cold snap (between the 15th and the 18th centuries).

The cyclic character of climate warming in the Arctic region has been observed both in the last 1,000 and 100 years. In the above cycles, the curves reflect the results of varved sediments (foliation) in the Izmenchivoe Lake (Severnaya Zemlya), and also in the other Arctic lakes. The speed of sediment accumulation is closely connected with the air temperature, especially during the summer period.[9] These curves make above century cycles (i.e., cycles covering several centuries) close to century (i.e., covering a whole century) and intra-century (i.e., covering periods shorter than a century) cycles distinguishable. The duration of the above century cycles ranges from 200 to 450 years. Shorter cycles (rhythms) last from 10–11 years up to 100 years. Rhythms of 30–40 years and 60–70 years are most clearly marked. Cycles covering several centuries and their phases are divided into warm and cold, respectively (Fig. 3.2).

Warm phases were present at the end of 10th–11th centuries, the 12th centuries, in the middle of the 14th ("the Epoch of Vikings"), the second half of the 15th, and the first half of 16th, 19th and the 20th centuries. Cold phases were present in the 9th, the beginning of the 10th, the end of the 11th, the beginning of the 12th, from the middle of the 14th to the middle of the 15th, and from the middle of the 16th to the beginning of the 19th century. In the coldest phase (Small Glacial Epochs), the temperature fell 1.0–1.5° compared to the modern average summer temperature in the month of July (about 1.5–2°). In the warmest phase (Epoch of Vikings), the temperature, on average, went down by 2°C and, in short periods, by 4–5°C.

Quasi-rhythmic fluctuations of climate parameters (mainly of temperature and precipitation) have appeared in the Arctic region over the last 100 years. This is illustrated well by indicators of average

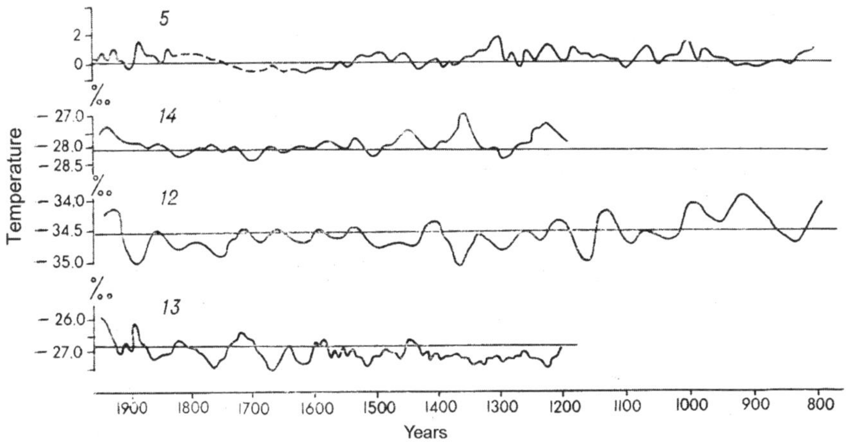

Fig. 3.2. Fluctuations of Indirect Indicators of Air Temperature for the Last Millennium.

Note: Change of a thickness of annual layers of varve of the lake Izmenchivoe (archipelago Severnaya Zemlya); it is constructed with application of the 10-year-old time filter (for the upper plot 5).

Source: Mihailovich, MV and BD Yur'evich (1991). The air temperature in the Holocene. In *Climatic conditions in the Arctic on the turn of the XX–XXI centuries*, Chapter 5, pp. 160–169. Leningrad: Hydrometizdat.

annual variability of air temperature in the western sector of the Arctic regions over the last 100 years (Fig. 3.3).

In five places — Arkhangelsk, Naryan-Mar, Amderma, Marresale, and Gyda — we see that at the beginning of the 20th century, the temperature was lower than the present one.[11] During the 1920–1940s, it rose by 2–3°C, then, during the 1950–1980s, it fell. Thus, the amplitude was almost 3°C. From the 1980s to now, there has been an obvious increase in average annual temperature. However, even in the first decade of the 21st century, the temperature has not exceeded the indicators of warming of 20–40 years.

Deviations from the standards of the average annual air temperature and precipitation over the last 70 years (from 1935 to 2003) in the sea basins of the AO are shown in Figs. 3.4 and 3.5.[12] In these two figures, for all basins as well as the land, one can see three periods: two warm periods (in the 1930–1940s and during the last 25–30 years)

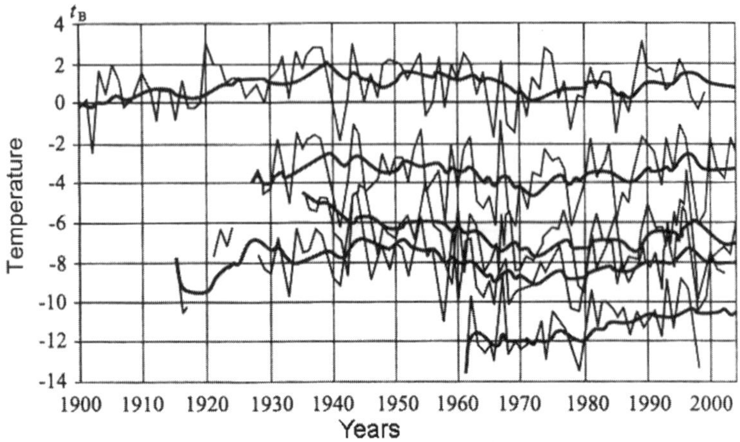

Fig. 3.3. Change of Average Annual Air Temperature in the (plots from top to down) Arkhangelsk, Naryan-Mar, Amderma, Marre-Sale, and Gyda Regions.[10]

Note: Thin lines — change of average annual temperatures, thick — moving average data on decades.

Source: Alexeevich, VA (2007). *The Geoecological condition of the Arctic coast of Russia and safety of nature management*, p. 585. *Moscow*: Geos.

and a cold period (in the 1950–1970s). Thus, the amplitude of the deviation from the norm for the western and central Arctic regions (basins of the Barentsev, Kara and Laptevyh) for the warm period of the 1930s–1940s was bigger compared to the last decades, i.e., the air temperature in the 1930–1940s, for the greatest part of the western and central Euroasian Arctic region, was higher compared with the last decades, not to mention the earlier episodes (an epoch of "Vikings", a climatic optimum of Holocene). The temperature trend for the last 70 years for these basins has been negative (Fig. 3.4).

In the East-Siberian Chukchi Sea and the Beaufort Sea, the amplitude of deviations from the norm in the 1930–1940s period was lower compared to that which occurred during the last decade, and the general temperature trend was positive (Fig. 3.5).

For precipitation, the scenario is completely different. For most basins, a subzero trend is characteristic in the examined period and only for the Barents Sea pool is it extremely positive (Fig. 3.5).

The information given above attempts to answer the question: What place does the warming of the last 30 years in the Arctic

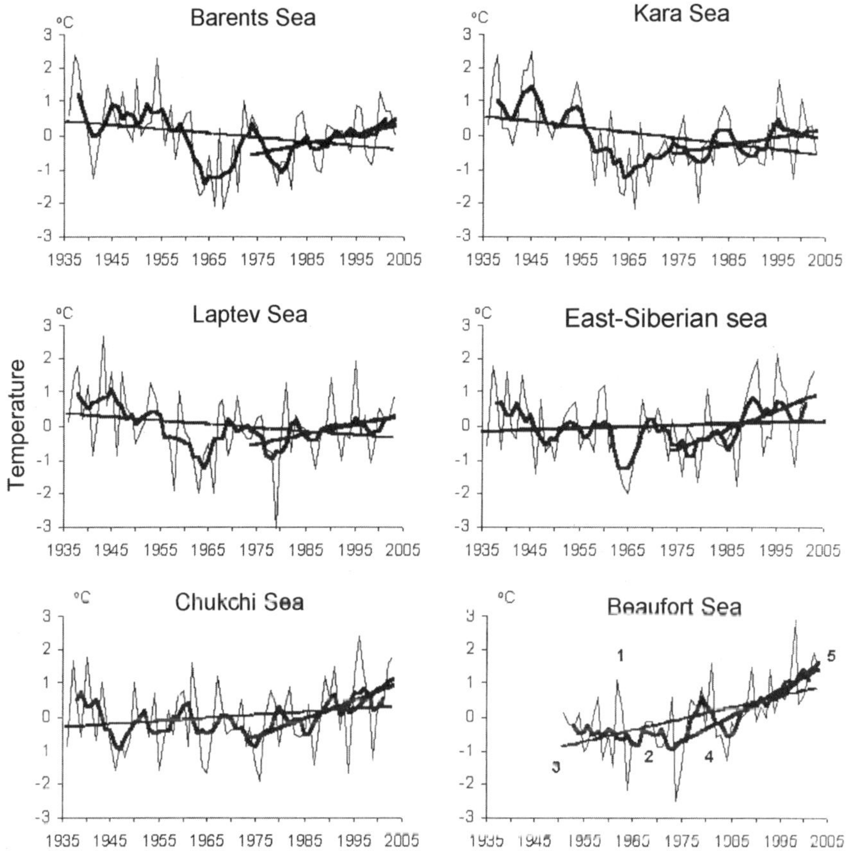

Fig. 3.4. Annual Changes of Normalized Anomalies of Average Annual Air Temperature (Celsius).

Notes: 1 — normalized anomalies; 2–5 — summer moving averages; 3, 4, 5 — the linear trend for the period of 1936–2003, 1974–2003.

Source: Ivanovich, AE and BN Nikolaevich (2008). Current changes of ground air temperature and atmospheric precipitations in Arctic. In *Problems of Arctic and Antarctic*, Vol. 446, pp. 31–40.

region occupy in relation to the previous climatic episodes? The answer can be only one: It is one of the of quasirhythmic climate fluctuation episodes, which have taken place over a long period of time and on the amplitude of the whole series of earlier warmings.

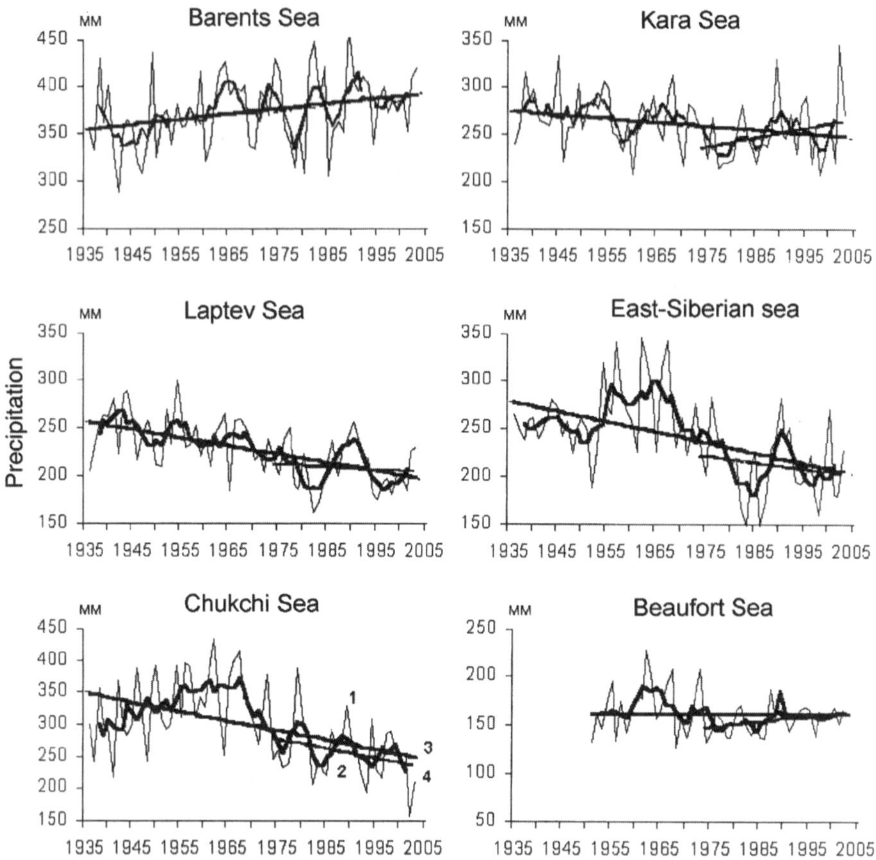

Fig. 3.5. Annual Change of Precipitation Total Amount (in mm).

Notes: 1 — precipitation total amount. 2–5 — moving average precipitation the total amount for summer, 3, 4 — linear trend for the period of 1936–2003, 1974–2003.

Source: Ivanovich, AE and BN Nikolaevich (2008). Current changes of ground air temperature and atmospheric precipitations in Arctic. In *Problems of Arctic and Antarctic*, Vol. 446, pp. 31–40.

What are the reasons for this warming? Unfortunately, there is no definite answer to this question. Over the last 15–20 years, one of the most popular hypotheses, especially in the mass media, has been the connection between global warming and a constant increase in atmospheric content of carbonic gas, methane and other greenhouse gases, caused by growing human activity (burning of fossil fuels, cutting down of trees, ocean pollution, etc.). According to the reports of the IPCC (2001, 2007), since the beginning of the Industrial Revolution, the content of carbonic gas in the atmosphere has increased by 35%, and the global average annual temperature has also increased, by 0.7–0.8°C.

According to the supporters of the "greenhouse effect" hypothesis, the further continuous increase of greenhouse gases will cause exponential logarithmic growth in the global air temperature, and for the Arctic region with its specificity of global processes, this growth should be 2–3 times higher. According to the various models, the global air temperature on different models will increase by the end of the 21st century by 1.4°C–5.8°C.[13] For the Arctic region, the highest temperature increase is expected: 3–5°C over land and up to 7°C over water; thus, winter temperatures should grow by 4–7°C over land and 7–10°C over ocean.[14] The amount of precipitation across the entire Arctic region will increase by 20%.[15] Reduction of the sea ice area in the AO will be more than 50%.[16]

There is also another group of researchers (including very prominent climatologists and oceanologists) both in Russia and abroad[17–20] who criticize both the "greenhouse effect" theory and the methodology of the climatic processes modelling currently used. These researchers see a connection between the contemporary climate warming and natural processes development. Among these processes, the main role is played by the gravitational processes of the solar system. They influence the speed of the rotations of the Earth, the angle of the slope of its axis and its orbit position, thereby, intensifying its irradiation.

According to the geophysical point of view, global warming has roots in space and this is the real reason for climate reorganization, whereas greenhouse atmospheric gases play only an insignificant part in the process.[21–23] The main cause is related to the solar-planetary processes. They include geological and geophysical processes on the Earth. The most significant ones are: (1) inversion of the sign on the geomagnetic field of the Earth (i.e., the change of poles) and decrease in the Earth's magnetic protection at the expense of a reduction in intensity of a geomagnetic bipole;[24] (2) the general direction of heat content in firm, liquid and gaseous phases of substance in the top cover of the Earth; (3) the growth of volcanic activities and seismic processes; (4) energy inflow linked to solar-terrestrial interrelations; and, (5) the enormous gain of electromagnetic energy of technogenic origin.

The direct consequence of these changes in the geological and geophysical environment are climatic processes of the "new generation" — disbandment of the general fields of the atmospheric pressure, changes in the temperature maps of ground atmosphere, and condition of the water-rotation.[25,26]

Quite probably, unprecedented capacity in the natural variations in the fields of atmospheric pressure (Arctic oscillation) arising in the Arctic region at the end the 1980s to the beginning of the 1990s strengthened the cyclonic activity and caused subsequent changes in the direction and force of winds, drift of sea ices, oceanic streams and increases in the river drain due to atmospheric precipitations. These were all the result of the development of geological and geophysical processes.

As for the reasons for the global warming over the last 30 years, the author of this chapter is inclined to consider that global warming has been the result of the development of natural processes connected with the gravitational influence of the solar system on the cover of the Earth. Only a combined interaction of galactic influences, star systems, and the solar system planets' orbiting around the sun can explain the rhythm in the change of seasons, the periods of warming, glacial and interglacial epochs and especially long galactic seasons. In recent years, for the purpose of revealing the influence of space factors on the climate, special attention has been given to the influence of the three largest planets of the solar system (Saturn, Jupiter, and Uranus) on solar activities, the speed of the Earth's rotation, change of gravity, and fluctuations in the rotation axis of the Earth's poles.[27]

Actually, over the last 400 years, connections to and oppositions from Jupiter and Saturn have occurred every 60 years (Table 3.1).

In years when connections occur on solar-longitude, close to the lowest average distance between the perihelion of the orbits of these

Table 3.1 Evaluation of Annual Anomalies of the Air Temperature and the Angular Rotation Speed of the Earth-Years of Connection and Opposition of Jupiter and Saturn.

The year of connection	**The anomaly of temperature**		**The angular speed of rotation of the Earth**	
	Year	**Maximum**	**Year**	**Maximum**
A. Years of connection of Jupiter and Saturn				
1581	1580	0.1		
1642	1641	0.1		
1702	1703	0.05	1700	124.4
1762	1764	0.32	1766	30.1
1822	1824	0.38	1820	186.9
1881	1880	0.1	1882	60.8
1940	1939	1.3	1933	38.2
2000	2000	1.8	2000	−83.8
B. Years of opposition of Jupiter and Saturn				
1611	1610	−0.78		
1672	1674	−0.62	1676	−57.9
1735	1734	−0.38	1734	29.5
1791	1785	−0.61	1792	−61.9
1851	1845	−0.45	1851	−60.2
1911	1905	−1	1910	−486.1
1971	1966	−1.3	1972	−370.9

Source: Alexandrovich, DM and BV Alexandrovich (2006). *The space, the planetary changes and the climate of Pole regions*, p. 358. Saint-Petersburg: Gidrometeoizdat.

planets (about 30°) and when they are closest to the sun, the resulting gravitational movement is the greatest, whereas at the opposition, it is minimal. Thus, in the years close to the connection, the annual air temperatures were extremely high, and the Earth's speed of rotation reached its maximum. In the years close to the opposition of Jupiter and Saturn, the annual air temperatures became extremely low, as well as the speed of the Earth's rotation. The temperature difference between these phenomena, for positive values, did not exceed 1–2°C, and, for negative values, 2–4°C.

Besides the 60-year- cycle, there have also been well expressed 11-, 20- and 80-, 90-year- cycles. The nature of the 11-year cycle is determined by the angular position of Saturn and Jupiter whose connection has occurred every 20 years. Eighty-and 90-year- cycles[28] are connected with the influence of current-generating forces of Uranus upon changes of solar activity; their cycle time around the sun is 84 years.

The powerful influence of the gravitational forces on the Earth cover can be observed today. For example, the recent approach of the moon towards the Earth has led to the occurrence of strong flows of water into the oceans and the Earth's crust. As a result, the majority of large ice fields in the AO have been destroyed. The Russian station NP-38 was an airfield without ice. Quite possibly, the powerful earthquake in Japan in March 2011 was also a result of the approach of the moon towards the Earth.

The possibility of considerable global warming caused by growth in greenhouse gas emissions due to anthropogenic activity is quite exaggerated because all of the natural change trends are not included in the so-called "greenhouse effect trends". According to some Russian researchers, methodological errors in the appendix of the greenhouse effect theory contradict the theory of climate and, to some degree, discredit all the ideas about the greenhouse effect theory.

The influence of contemporary warming on the Arctic environment has not only positive consequences, but negative ones. The migration of some plant and animal species to the North, including those of trade value (e.g., elk, bear and fox), has begun. At the same time, displacement of vegetative sub-bands of plains to the North is not observed. Due to atmospheric precipitation reduction on the largest part of the territory, many lakes and rivers have become shallow, affecting the fisheries considerably.

Significant and distinctive changes have taken place in the cryosphere. During the warming period, land glaciers lost weight and decreased in size. Glaciological research made during the International Polar Year (2007–2008) showed that the glaciers of the Arctic from 1961 to 2005 lost 660.1 km^3 of their ice volume, which is roughly equivalent to the increase in the world ocean level by 0.18 mm/year.[29] This loss in the ice volume occurred not only due to the so-called climatic factor, but, to a larger extent, to the dynamic factor (drainage of the ice into the sea and iceberg formation). Thus, for instance, iceberg drainage accounts for more than 50% of the Franz Joseph Glacier's loss of mass.[30]

In the Russian Arctic, from 1952 to 2011, ice losses in the main centre of glaciations (the Franz Joseph Land, Severnaya Zemlya, and Novaya Zemlya) accounted for approximately 1.6% of the starting mass (maximum being 2.8% for the Franz Joseph Land and minimum being 0.4% for Severnaya Zemlya).[31] The permafrost in several areas resulted in small degradation at the cost of growth in the season-thawed layer and a rise in the temperature of the top horizons. In other areas, for example Yakutia, due to a decrease in deposits during the winter period, there was more intensive permafrost of soil and thawing signs were not observed.

One of the key indicators of climatic change in the Arctic region is the condition of sea ice in the AO. The changes in the regime and area of distribution of ice in the AO are of great importance with respect to the development of Arctic sea transport systems and the exploitation of minerals in the Arctic shelf. Based on satellite data, the area of September sea ice in the Northern Hemisphere gradually decreased from the 1970s to 2007. In 2007, the smallest area of sea ice at the ocean for the entire period of satellite data collection was observed. The area was 4.3 million km^2 whereas it was equal to 7.5 million km^2 in 1979.[32]

Such dynamics in the ice cover prompted many researchers to consider that in future, the area of sea ice in the AO would decrease gradually. However, only for the winter period of 2008–2009 (a severe winter), did the area of sea ice, as well as the area of distribution of old ice grow again by 611 thousand km.^{2} This corresponds to the reduction in the area of ice cover over all the seas over the Eurasian shelf during the 21st century.[33]

Fig. 3.6 shows the dynamics in the area of sea ice in the AO from 2007 to 2010. Produced by Russia's AARI, this map shows that the ice area essentially grew in the Laptev Sea, East-Siberian and Chukchi up to the beginning of navigation in 2008–2010; thus all the main passages located on a line in the Northern Sea Route (NSR)

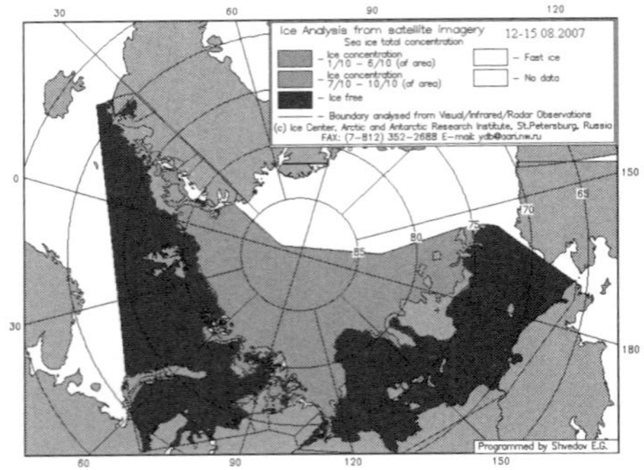

Arctic ice map for 12-15.08.2007

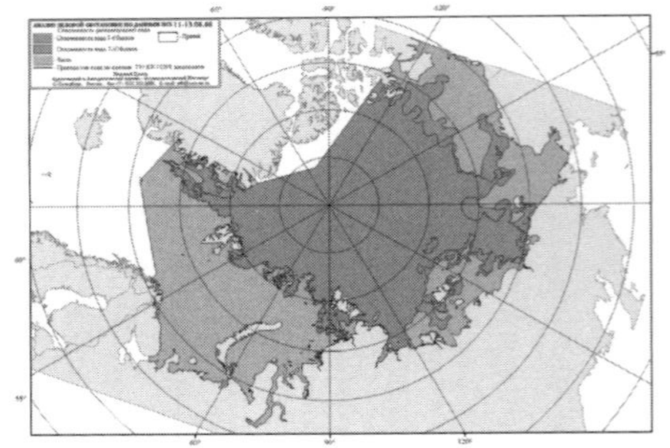

Arctic ice map for 11-13.08.2008

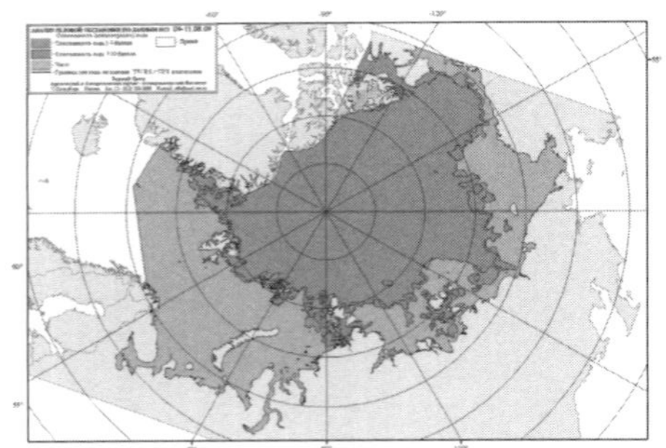

Arctic ice map for 9-11.08.2009

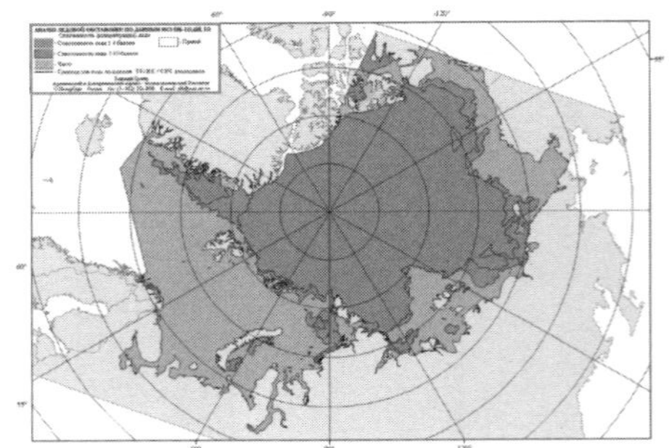

Arctic ice map for 8-10.08.2010

Fig. 3.6. Dynamics of the Area of Sea Ice in the AO 2007–2010.

Source: The Arctic and Antarctic Research Institute (AARI) (2010). *Survey of Hydrometeorological Processes in the Northern Ocean*, Appendix 1 Figs. 2.26–2.29. Saint-Petersburg: GNCRF. www.aari.ru

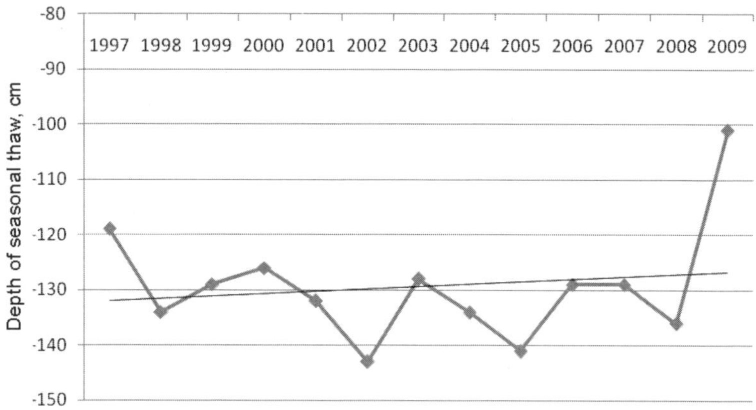

Fig. 3.7. Thickness Change of a Seasonally-Thawed Layer in Nadim Station, 1997–2009.
Note: In this site, decreasing of thaw depth of seasonally-thawed layer is observed. It is possible to explain that winters 2009 and 2010 in Yamal were very severe, and the temperature fell down to −50°C.

Source: Figure was created by the author based on the available data at CALM SUMMARY DATA TABLE. WEST SIBERIA (Nadym) R1. Circumpolar Active-Layer Monitoring (CALM) with this URL (http://www.udel.edu/Geography/calm/data/data-links.html).

were blocked by ice, namely the Kara Gate, Vilnitsky, Shokalsky, Dmitry Laptevyh, and Longa Straits.

According to the data collected by the Circumpolar Active Layer Monitoring, the power of the seasonally-thawed layer of the permafrost soil, the same process of renewal of the permafrost, became apparent in some areas of the Russian Arctic and in North America. This is well evident in Figs. 3.7, 3.8 and 3.9.[34]

3.3. ANTHROPOGENIC INFLUENCE

Analysis of the experience of the previous stage of natural resource potential in the Arctic zone shows that the negative consequences began to affect the environment and the native population. Among all the consequences of economic, military and other anthropogenic activities, the most dangerous for the Arctic environment is the constantly growing pollution.

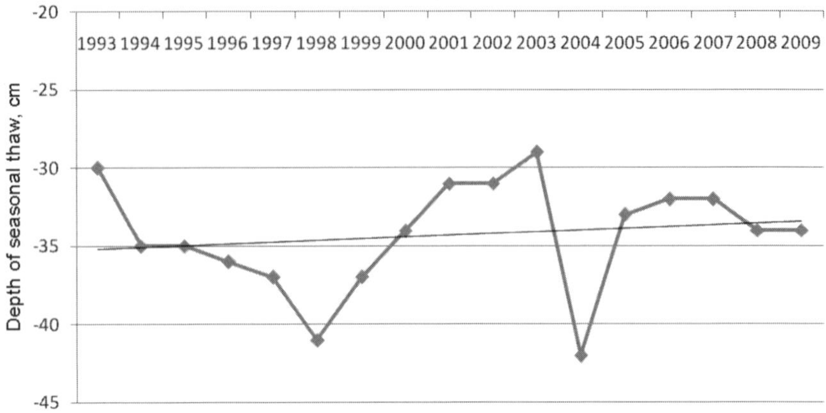

Fig. 3.8. Thickness Change of a Seasonally-Thawed Layer in Barrow Station (Alaska), 1997–2009.

Note: Analysis of the trend line indicates that there is no change in the depth of the thaw.

Source: Figure was created by the author based on the available data at CALM SUMMARY DATA TABLE. WEST SIBERIA (Nadym) R1. Circumpolar Active-Layer Monitoring (CALM) with this URL (http://www.udel.edu/Geography/calm/data/data-links.html).

The sources of pollution are local industrial materials and polluting substances (PS) from outside of the Arctic which appear as a result of transboundary transfer over the water and air. Such materials are transferred to the Arctic or move within the region through different means as detailed in the following sections.

3.3.1. *Atmospheric transference*

Atmospheric transference of PS to the Arctic is subdivided into two types: single-phase and repeated. Single-phase transference is the transference of chemical compounds, which get into the atmosphere and their precipitation leads to their undersurface accumulation where they stay. This type of transference is typical of acids, heavy metals, excluding mercury, and non-evaporating elements precipitating on solid organic compounds like benzpyrene and particles of black coal. Polluting substances, once transferred, are characterized by quite a long existence in the atmosphere in winter (approximately 20–30 days) and a short existence in summer (three to seven days).

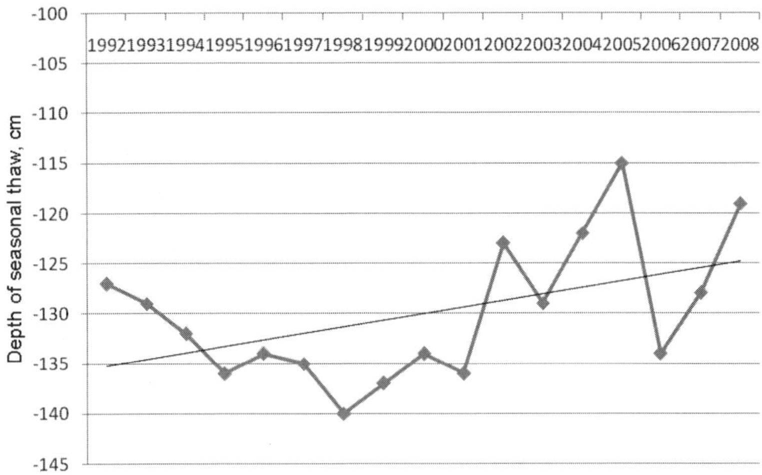

Fig. 3.9. Thickness Change of a Seasonally-Thawed Layer in Reindeer Deport Station (Canada), 1997–2009.

Note: It is clear that there is a decreasing trend in the depth of the permafrost layer in this picture.

Source: Figure was created by the author based on the available data at CALM SUMMARY DATA TABLE. WEST SIBERIA (Nadym) R1. Circumpolar Active Layer Monitoring (CALM) with this URL (http://www.udel.edu/Geography/calm/data/data-links.html).

They are also characterized by the predominance of the transference from the South (Eurasia) to the North (Arctic) in winter. Transference of PS to the Arctic is determined by the different character of the general circulation of large amounts of air over the Arctic region in winter and summer. In winter, the air is transferred by the strong Siberian anticyclone basically from Eurasia to the Arctic and farther to North America or to areas of high cyclone activity, i.e., the Aleutian Islands and Southern Greenland. As a result, in winter, the emission of PS from the sources in Eurasia and partly in the Russian North dominates the pollution of the environment in the Arctic, while, in summer, the main source of pollution is the inflow of PS emitted in Europe via the North Atlantic, and in North America via the Northern Pacific.

Thus, the emission of PS from anthropogenic sources is the main reason for the pollution of the Arctic atmospheric air in winter at an altitude of 2–3 km. The emissions of PS from Eurasian sources accounts for more than a half of the winter pollution of the Arctic atmosphere. The main sources of pollution are considered to be the Urals, Kola Peninsula, and Norilsk industrial regions, and industrially-developed countries of central Europe. Emissions from the Urals and Norilsk are the main sources of air pollution on the Taimyr Peninsula, the Chukchi Peninsula and other regions of the Eastern part of the Russian Arctic, Alaska and the Western part of the Canadian Arctic and Northern Scandinavia. Briefly, in winter, the transference of PS from Russian sources predominates, and in summer that from the European and North-American sources predominates.

3.3.2. *Water transference*

Drainage of the big Arctic rivers is one of the most significant means for transference of pollutants to the AO. Annual drainage of dissolved substances into the closing sites of the rivers in the North and Siberia is estimated in the hydrochemical and hydrological data from the State Service for Observation and Control (Roshydromet). Normally the drainage of organic substances is calculated according to the chemical classifier, ammonium, nitrate, nitrite nitrogen, general phosphorus, general iron, silicon, petrochemical products and some chlorine organic pesticides.

A fundamental attempt at estimating the changeability of water and chemical drainage in the Russian Arctic as a result of anthropogenic activity was made by Nikanorov, Ivanov, and Bryzgalo.[35] According to their data, the biggest annual long-term water drainage in the AO belongs to the Yenisei River (621 cubic km), the Lena River (542 cubic km), and the Ob River (537 cubic km). As a result of the huge water content, these rivers have the lowest, compared to other big rivers, long-term changeability in the water drainage of the undercurrent ($C_{\mathrm{v}} = 0.12 - 0.15$). A slight increase in the annual water drainage has recently been recorded in most rivers excluding the Ob, Lena, and Kolyma. The Pechora River has recorded the highest increase in the annual water drainage. On the whole, the water drainage from all the rivers varies from year to year, but annual variations are loosely connected. Synchronism of the annual water drainage variations is especially noticeable in the Yana and Indigirka Rivers. Antisynchronism is characteristic of the Lena and Kolyma Rivers.

Annual drainage has two main seasons. Most occurs during the warm seasons (from May to October), i.e., spring and summer seasons. The watersheds of the rivers flowing into the White, Barents and Kara Seas vary from 79.5% to 85.2%, and those of the rivers flowing into the Laptev and East Siberian increase from 92.8% to 98.8%.[36]

The anthropogenic factor is the key factor in the transformation of the hydrochemical mode of many Arctic rivers. The consequences of this influence reveal themselves in the extent of the range of concentrations of the main ions, biogenic components and dominating PS in the aqueous medium together with the general trend of increase in the reiteration of high multiplicity beyond maximum permissible ecological concentrations (MPEC) and maximum permissible concentrations (MPC). It was under this influence that in the last 20 years, the tendency of mineral forms of nitrogen's and phosphorus's periodical accumulation in aqueous medium has clearly appeared.[37] The accumulation has reached a concentration dozens and hundreds times beyond the MPEC for eutrophic basins. The same situation is recorded in aqueous medium for dominating polluting substances, including the following: petroleum hydrocarbons, phenols, copper, zinc, iron compounds and easily oxidizable organic compounds.[38]

Analysis of the long-term hydrobiological mode observations confirm the following conclusion. A threat of regression and degradation of ecosystems on estuary areas of the rivers in the Russian Arctic is a possible and dangerous consequence of emergency ecological situations in conditions when water objects undergo long-term (chronic) anthropogenic influence, as a result of which a new "anthropogenically changed" ecological state is formed. This development results not only in the change of ecosystems' status, but in the decrease of their ecological capacity.[39] Further expansion and development of hot spots may lead to negative changes and ecological catastrophe.

3.3.3. *Biological transference*

The existing data and information available to the federal and regional authorities dealing with environmental protection and health care concerning the sources of pollution with steady organic compounds (SOC) do not reflect the situation in the Russian Arctic adequately. The main sources of nutrition for most of the native population are venison, the fat of sea animals and fish. Short food-chain lichen, reindeer, or water reservoirs (consumed by plankton, fish and human beings) provide for direct entry of PS, heavy metals, chlororganic compounds, and long-lived radio nuclides into the human organism. A daily entry of radio nuclides reaches 2960–3700 Bq[40] in the Arctic region, i.e., ten times more than that for the citizens of Central Russia.[45] Recent research has also recorded the influence of some local sources unknown before. For instance, there is evidence of the emission of certain steady organic compounds (SOC), such as dichlorodiphenyltrichloroethane (DDT) and Phencyclidine (PCP) into the environment.[42]

In general, the levels of concentration of PCP in physical and biotical components of the Russian Arctic environment are moderate compared to the other regions of the Arctic. This result has occurred due to the considerable reduction in PCP entry into the native population, without interfering with their traditional way of life and cultural identity, with the help of protection and reconstruction measures, including improvement in the sanitary conditions of the natives and their following recommendations regarding house maintenance and nutrition.

At the same time, the levels of exposure in the Russian Arctic to PCP and to hexachlorobenzene and hexachlorohexamethylene, in particular, and, in some cases, to DDT and PCP, are the highest among all the Arctic regions.[43] In some cases, the exposure exceeds the levels accepted internationally and in the zones of ecological disaster, such as the Aral Sea region due to the long-term use of pesticides. In the investigated regions of the Russian Arctic, almost every family consumes a considerable amount of traditional food. Low-income families consume traditional food rich in locally produced lipids. As a consequence, they find themselves facing the highest risk of exposure to SOC.

The highest levels of exposure to SOC and the health risks resulting from them have been recorded on the coast of the Chukchi Peninsula, where the traditional diet is based on the consumption of sea mammals and fish. This corresponds to information about the diets of the Greenlandic and coastal Canadian population. The concentration of SOC in blood and breast milk in the mentioned cases correlates with the level of SOC among the female population of the Chukchi Autonomous District.[44] There are serious reasons to consider that a number of diseases (oncological, allergy, hematological, and hereditary pathology) result from the unfavourable ecological conditions of the territories.

The environmental pollution influences the soil biota and small animal species (lemmings, voles etc.), which are at the foot of the trophic pyramid. A considerable decrease in the number of many small rodents affects the fodder supplies to Arctic foxes and other predators, including those with feathers. The consumption by the predators of the rodents that died from chemical poisoning influences the reproductive functions of the predators considerably.

In those parts of the Arctic with active economies, the soil-vegetation cover is damaged considerably as a result of: pollution, mechanical influence, the construction of roads, settlements, piping, and industrial facilitities, the warehousing of secondary rock for disposal, fires, and reindeer overgrazing. Regions with a long history of economic activities (e.g., Norilsk, Kola, and Vorkuta) are the most-affected in this respect. In these regions, the soil and vegetation cover has been destroyed within a hundred kilometres' radius around industrial enterprises and the scenery has acquired the features of an Arctic waste land. Domestic reindeer pastures have also deteriorated. Only 35% out of more than 330 million hectares of pastures remain in acceptable condition.[45] In the last few years, degradation of the soil cover has been especially severe near the oil production sites in the Nenets and Yamal-Nenets autonomous regions.

Anthropogenic factors also considerably influence the condition of Arctic animals, vegetation populations, and biocenoses. The natural habitations change, the resources disappear and the environment is polluted. Disappearance of their food resources is the most dangerous development for the preservation of especially valuable species. It is manifested in the uncontrolled fishing and hunting of rare species done by both private and state companies and by numerous individual poachers. Poaching in the Arctic is as profitable as, if not more profitable than, the enterprise sector of the local economy.

Environmental pollution, expansion of the anthropogenic landscapes, and weakly regulated operation of biological resources unfavourably affect the Arctic ecosystems' stability, its biological diversity, its population's health, and its economic position leading to occurrence of the so-called "hot spots". For example, 140 "hot spots" appeared in the Arctic Zone of the Russian Federation and Northern territorial formations bordering on the South in 2004.[46] The greatest number is observed on the European part of the territory. Their formation is connected with the activities of large thermal power stations, energy transportation (including via pipelines), pulp-and-paper manufacturing, the housing and communal services of cities and settlements, mountain-concentrating enterprises, metallurgical industrial complexes, and military units, among other factors.

The activities causing the fastest progress in the formation of "hot spots" in the vast territories of the North of Russia are extraction and transportation of oil and gas which creates not only powerful anthropotechnical influences on the land ecosystems, but has evidently pressured the coastal ecosystems through the drainage system. The general loss of oil hydrocarbons during extraction and oil transportation affecting the Arctic environment is estimated to be in the range of 3% to 7%.[47] The anthropogenic streams coming to the environment are multi-component, containing such substances as oil and oil products (e.g., methane, oil, aromatic and other groups of hydrocarbons and substances available in them such as pitches, mercaptans, mercury, and arsenic), condensates, hydrocarbonic gases, poly-cyclic aromatic hydrocarbons, oxides of sulfur, nitrogen, and carbon. Almost all substances, which are ingressed in technogenic streams formed in the enterprises, are active geochemically, very mobile and capable of cooperating actively with natural systems, are frequently highly toxic, and, therefore, dangerous for biocenoses. The concentration of oil and gas extraction in certain areas has led to the fact that the process of polluting the natural waters has acquired a regional character, covering territories of large oil and gas-producing provinces, especially dangerous in the northern territories of Russia.

Oil and gas extraction also pollutes the atmosphere. For example, annual gas flaring in Western Siberia is up to 10^{15} cubic meters and up to 10^{7} cubic meters in Timano-Pechora Province, thereby polluting the environment with oxides of carbon, nitrogen, and soot.[48]

The concentration in certain "hot spots" leads to the formation of impact areas. These refer to an area where negative changes in the environment occur as a result of anthropogenic influence, which have led to the occurrence and development of extreme ecological situations. Impact areas have the following characteristics:

Transformation of the natural geochemical background
Atmospheric pollution
Pollution of natural waters
Degradation of a soil/vegetative groundcover
Introduction of PS in food chains
Increase in the frequency of certain diseases among the populations.

For the affected Arctic territory, there is a distribution of local impact areas associated, as a rule, with certain economic profiling closely connected with the location of mineral deposits or other natural resources. The distribution of impact areas is extremely irregular. The largest number is concentrated in the Western part of the Eurasian North on the reservoirs of the rivers flowing through the biggest industrially-developed territories and the populated areas of Murmansk, Arkhangelsk areas, the Republic of Komi, Nenetskij, the Yamal-Nenetskij districts, and, to some extent, the Tajmyrsky Municipal District.

Figure 3.10 shows the areas of intensive economic activity in the Russian Arctic region marked with numbers and having the following characteristics: 1–9 the basic impact areas (anthropogenous impact on estuarine ecosystems of Northern and Siberian rivers consisting of the river basins (1-Kola; 2-Severodvinsk; 3-Timano-Pechora; 4-Vorkuta; 5-Sredne-Obsky; 6-Pur-Nadym; 7-Norilsk; 8Jano-Indigirsky; 9-Valkumsjsky); 10-the impact areas' borders; 11-hot spots with very critical (crisis) ecological conditions; 12-hot spots

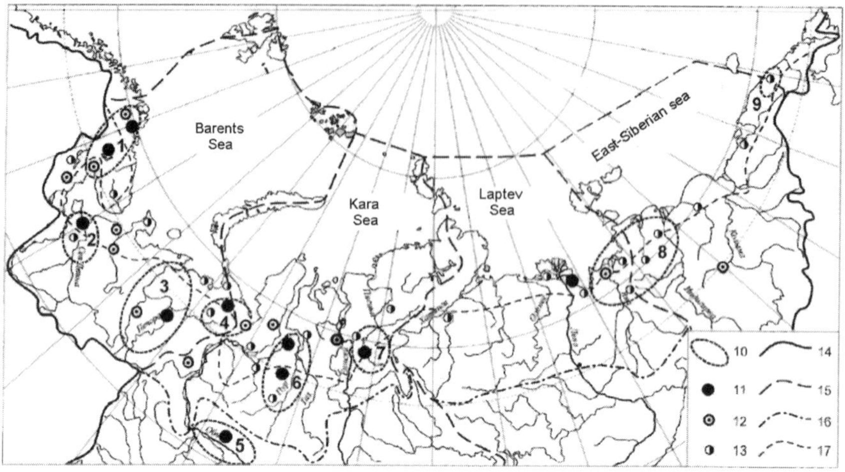

Fig. 3.10. Main Impact Areas within Land Catchment Areas of the North and the Arctic of Russia.
Source: Alexandrovna, BV, VV Ivanov, VY Zamyatin and VM Makeev (2004). *Emergency Ecological Situations in the Russian North*, p. 22. Saint-Petersburg: State Polar Academy.

with critical ecological conditions; 13-hot spots with moderately critical ecological situations; 14-borders of the AO catchment areas; 15-waterecological border of the AO region; 16-catchment areas of the AO seas; 17-border by the river closures).

3.4. CONCLUSION

Predicting climate change and its influence on natural and social processes over the near future is a rather difficult task, even if one uses the most modern mathematical models. The principal difficulty is that climate change has too many contributing factors and it is difficult to distinguish their individual contributions.

The head of the Hydrometeorological Service of Russia, A.E. Frolov and his colleague, the well-known climatologist, V. Kattsov wrote about this.[49] The most acceptable approach is an analog or extrapolating approach, which permits extrapolating the tendency towards climate change over a sufficiently long observation period. Based on such an approach in 2005, the Federal Hydrometeorology and Environmental Monitoring Service of Russia published *The Strategic Forecast of Climate Changes of the Russian Federation and Maintenance of National Safety for the Period of 2010–2015.* In this publication, the AARI, drawing on its many decades of studying ice conditions in the AO upon which it bases its annual forecasts of navigating conditions in the NSR, forecast possible changes in the Arctic seas' ice conditions for the following 10 years.[50] Accordingly, it was expected that the ice conditions would be formed on the reduced background of ice by the preservation of an existing background of ice and hydro weather conditions. Thus, the formation of difficult and very difficult ice conditions are quite possible (repeatability of 10–20%) in the Straits of Vilkitsky, Shokalsky, Dmirija Laptevyh, Sannikova and Long, limiting navigation without icebreakers in the NSR.[51] The duration of the period of navigation on the NSR without ice breakers when there are difficult ice conditions is reduced to 10–15 days a year (compared to the ordinary two-month period as expected). Figure 3.10 illustrates the reliability of the forecasts for the ice cover conditions from 2007 to 2010. In the same forecast, it is emphasized that the duration of the ice period at the NSR to the east from the Kara Gate Strait will exceed six months per year and consequently the basis for Russia's full sovereignty over the NSR will remain in accordance with Article 234 of the United Nations Convention on the Law of the Sea.[52]

To forecast ice conditions in the Arctic seas over the near term, it is most reasonable to rely not on the mathematical models, but on the huge experience of the experts at AARI. Modern techniques are usually deployed, including satellite information and mathematical modelling of natural processes, to forecast the long-term atmospheric and oceanic processes, as well as an ice regime. In addition, other revealed regularities in the development of natural processes are considered, particularly, the cyclic character of fluctuations in the ice, caused by rhythmically repetitive solar-planetary processes, thus there is a need for both traditional techniques as well as the newly-developed numerical dynamic-thermodynamic models. Another recent forecast holds that, even within the limits of climatic scenario A1B of the ECHAMS/MP1-OM, NAD CM3 and CCSR/NIES/FRCC GC models,[53] by the end of the 21st century, old sea ice will remain in the Arctic basin.[54,55]

Of key importance is the changing environment and ecological security in the context of global warming. As stated earlier, the current warming in the Arctic is considered a unique event. However, the amplitude of fluctuations in temperature does not go beyond the limits of natural fluctuations, which have taken place for almost the entire Holocenic period (about 10,000 years). Moreover, one can absolutely declare that the whole series of early warming periods greatly surpassed the current warming.

Very significantly, — the warming in the Arctic has not occurred everywhere. There are areas where the air temperature has not risen, but has in fact gone down. Moreover, this different temperature direction can be observed within the limits of one territorial region, for example, the Chukchi Peninsular region whose eastern part has experienced warming while its Western part has faced a cold snap.[56]

In short, to compare the anthropogenic and natural factors from the viewpoint of their unfavourable effects on the changes of the Arctic environment and human health, it is the anthropogenic factor that is the most dangerous. However, climatic changes have a cyclic character caused to a large extent by the different processes in nature. Consequently, climatic developments in the near future will most likely be in the form of further dropping of temperatures and consequently the growth of glaciers, increases in sea ice areas and permafrost thickness, decreases in the seasonal thawing depth and iceless navigation periods in the NSR, and the development of other processes characteristic of cold zones. Nevertheless, the exploration for and export of mineral and energy resources in the Arctic region, as well as efforts towards sustainable social and economic development and navigation in the NSR will continue notwithstanding the climate change and ice cover conditions in the region.

BIBLIOGRAPHY

Annual Report of the Expert Council on the Arctic (2010). M. Publishing House of the Council of the Federation.

Alexandrovich, DA and BV Alexandrovich (2006). *The Space, the Planetary Changes and the Climate of Pole Regions.* Saint-Petersburg: *Gidrometeoizdat.*

Alexandrovna, BV, VV Ivanov, VY Zamyatin and VM Makeev (2004). *Emergency Ecological Situations in the Russian North.* St. Petersburg: State Polar Academy.

Alexeeyskiy, NI (ed.) (2007). *The Geoecological Condition of the Arctic Coast of Russia and Safety of Nature Management.* Moscow: *Geos.*

Bedritskiv, AI (ed.) (2005). *The Strategic Forecasts of the Climate Changes of the Russian Federation for the Period till 2010–2015 and their Influence on Branches of Economy of Russia.* Rosgidromet.

Georgievna, TY (2003). The Account of Losses at Oil Recovery. The Dissertation Abstract of Candidate's Thesis in Economy, St. Petersburg.

Grigorievich, TB (2009). *The Pulsation of the Earth and Cycles of the Geodynamic Activity in the Streams of the Space Plasma.* St. Petersburg: *MANEB.*

IPCC (2007). *Climate Change 2007: The Physical Basis.* Contribution of Working Group to the Fourth Assessment Report of the Intergovermental Panel on Climate Change. Cambridge: Cambridge University Press.

Ivanovich, AE and BN Nikolaevich (2008). Current changes of ground air temperature and atmospheric precipitations in Arctic. In *Problems of Arctic and Antarctic*, Vol. 446, pp. 31–40.

Maximovich, NA, IV Vladimirovich and BV Alexandrovna (2007). *Rivers of the Russian Arctic in Modern Conditions of Anthropogenic Influence.* Rostov-on-Don: NOK.

Meehl, GA and TF Stocker. *Global Climate Projections.*

Mihailovich, MV *et al.* (2003). Vegetation and climate of the New Siberian islands for the post 15,000 years. *Arctic, Antarctic and Alpine Research*, 35(1), 56–66.

Mihailovich, MV (2011). The long-term changes of a thermal regime of Arctic. *Izvestia of Russian Geographical Society*, 1.

Mihailovich, KV (2011). From the international polar year to the international polar decade. *Arctic, Ecology and Economy*, 1.

Mihailovich, MV and BD Yur'evich (1991). The air temperature in Holocene. In *Climatic Conditions in the Arctic on the Turn of the XXXXI Centuries.* Chapter 5. Leningrad: Hydrometizdat.

Mihailovich, MV, BD Yur'evich and KA Vladimirovich (1983). The method of research of varve sediments of contemporary periglacial lakes for the purpose of reconstruction of paleoclimate. *The History of Lakes in the USSR.* Thesis of the report of VI all-USSR conference (Tallinn), 1.

Nikolaevich, DA (2007). Interview to the correspondent of RIA "Siberia." 4 October. http://ria-sibir/ru/viewnews/20501.html

Panel on Climate Change (2003). The survey report of the program *Estimations of Influence of the Climate Changes in Arctic Region (ACIA)*. Cambridge University.

Panteleimonovich, BY (1998). Gravitation, climate and weather. *Vestnik* (St. Petersburg State University), Series 7, 7(7).

Petrovich, MA, SS Viktorovich and DD Mihailovich (2010). The modelling estimation of ice conditions in the Arctic basin in the end of XX and the XXI-st centuries. The thesis of reports of final session of Academic council AARI by results of the work in 2010.

Sergeevich, MA (2000). *The Influence of Planets on the Earth Climate.* Moscow: Nauchnyj mir, MGU.

Sergeevich, FA and K Vladimi (2009). Predicting Arctic climate: Knowledge gaps and uncertainties. *Climate Change and Arctic Sustainable Development*, UNESCO.

The Strategy of Development of the Arctic Zone in the Russian Federation and Safeguarding of National Security in the Period up to the 2020 Year (2010). Ministry of Regional Development of the Russian Federation. Moscow.

Yakovlevich, KK (1992). *The Global Climate.* St. Petersburg: Nauka.

Yevgenievich, FI, GZ Markovich *et al.* (2008). Sixty-year cyclic changes in the climate of the polar regions. *Materials of Glaciological Research* (Moscow), No. 105.

ENDNOTES

[1] *The Strategy of Development of the Arctic Zone in the Russian Federation and Safeguarding of National Security in the Period up to the 2020 year* (2010). Ministry of Regional Development of the Russian Federation. Moscow.

[2] IPCC (2007). *Climate Change 2007: The Physical Basis*, Contribution of Working Group to the Fourth Assessment Report of the Intergovernmental Panel on climate change. Cambridge: Cambridge University Press.

[3] Bedritskiv, AI (ed.) (2005). *The Strategic Forecasts of the Climate Changes of the Russian Federation for the Period till 2010–2015 and their Influence on Branches of Economy of Russia*, p. 28. Rosgidromet.

[4] Mihailovich, MV *et al.* (2003). Vegetation and climate of the New Siberian islands for the post 15000 years. *Arctic, Antarctic and Alpine Research*, 35(1), 56–66.

[5] Mihailovich, MV (2011). The long-term changes of a thermal regime of Arctic. *Izvestia of Russian Geographical Society* , 1, 28–35.

6 Mihailovich, MV, BD Yur'evich (1991). The air temperature in Holocene. *Climatic Conditions in the Arctic on the Turn of the XX–XXI Centuries*, Chapter 5, pp. 160–169. Leningrad: Hydrometizdat.

7 Thermokarst — negative forms of relief.

8 Thermoabrasia — destruction of sea shores built from permafrost.

9 Mihailovich, MV, BD Yur'evich and KA Vladimirovich (1983). The method of research of varve sediments of contemporary periglacial lakes for the purpose of reconstruction of paleoclimate. *The History of Lakes in the USSR*, Thesis of the report of VI all-USSR conference (Tallinn), 1, 124–125.

10 *Ibid.*

11 Alexeevskiv, NI (ed.) (2007). The geoecological condition of the Arctic coast of Russia and safety of nature management, p. 585. Moscow: *Geos.*

12 Ivanovich, AE and BN Nikolaevich (2008). Current changes of ground air temperature and atmospheric precipitations in Arctic. In *Problems of Arctic and Antarctic*, Vol. 446, pp. 31–40.

13 Panel on Climate Change (2003). The survey report of the program *Estimations of Influence of the Climate Changes in Arctic Region* (*ACIA*). Cambridge: Cambridge University.

14 *Ibid.*

15 *Ibid.*

16 *Ibid.*

17 Panteleimonovich, BY (1998). Gravitation, climate and weather. *Vestnik* (St. Petersburg State university), series 7, 7(7), 87–91.

18 Yakovlevich, KK (1992). *The Global Climate*, p. 359. St. Petersburg: Nauka.

19 Sergeevich, MA (2000). *The Influence of Planets on the Earth Climate*, pp. 122–128. Moscow: Nauchnyj mir, MGU.

20 Alexandrovich, DA and BV Alexandrovich (2006). *The Space, the Planetary Changes and the Climate of Pole Regions*, p. 358. St. Petersburg: Gidrometeoizdat.

21 Nikolaevich, DA (2007). Interview to the correspondent of RIA "Siberia". 4 October. http://ria-sibir/ru/viewnews/20501.html

22 Sergeevich, MA (2000). *The Influence of Planets on the Earth Climate*, pp. 122–128. Moscow: Nauchnyj mir, MGU, *op. cit.*

23 Grigorievich, TB (2009). *The Pulsation of the Earth and Cycles of the Geodynamic Activity in the Streams of the Space Plasma*, p. 319. St. Petersburg: MANEB.

24 A magnetic dipole is a closed circulation of electric current. The Earth's magnetic field, approximated as a geomagnetic dipole. However, the "N" and "S" (North and South) poles are labelled here *Geographically*, which is the opposite of the convention for labelling the poles of a magnetic dipole moment

25 Nikolaevich, DA (2007). Interview to the correspondent of RIA "Siberia". 4 October. http://ria-sibir/ru/viewnews/20501.html, *op. cit.*

26 Grigorievich, TB (2009). *The Pulsation of the Earth and Cycles of the Geodynamic Activity in the Streams of the Space Plasma*, p. 319. St. Petersburg: MANEB. *op. cit.*

27 Alexandrovich, DA and BV Alexandrovich (2006). *The Space, the Planetary Changes and the Climate of Pole Regions*, p. 358. St. Petersburg: *Gidrometeoizdat.*

28 *Ibid.*

29 Mihailovich, KV (2011). From the international polar year to the international polar decade. *Arctic, Ecology and Economy*, 1, 4–19.

30 *Ibid.*

31 *Ibid.*

32 Yevgenievich, FI, GZ Markovich *et al.* (2008). Sixty-year cyclic changes in the climate of the polar regions. *Materials of Glaciological Research* (Moscow), 105, 158–165.

33 *Ibid.*

34 Circumarctic Active-layer Monitoring (CALM). http://www.udel.edu/Geography/calm

35 Maximovich, NA, IV Vladimirovich and BV Alexandrovna (2007). *Rivers of the Russian Arctic in Modern Conditions of Anthropogenic Influence*, p. 280. Rostov-on-Don: NOK.

36 *Ibid.*

37 *Ibid.*

38 *Ibid.*

39 *Ibid.*

40 Bq — Becquerel, unit of measuring radionuclide content.

41 *Annual Report of the Expert Council on the Arctic* (2010). M. Publishing house of the Council of the Federation.

42 DDT — dichlorodiphenyltrichloroethane, SOC — steady organic compounds, PCP — phencyclidine Bryzgalo.

43 Alexandrovna, V, VV Ivanov, VY Zamyatin and VM Makeev (2004). *Emergency Ecological Situations in the Russian North*, pp. 33–34. St. Petersburg: State Polar Academy.

44 *Ibid.*

45 *Annual Report of the Expert Council on the Arctic* (2010). M.: Publishing house of the Council of the Federation.

46 Alexandrovna, BV, VV Ivanov, VY Zamyatin and VM Makeev. *Emergency Ecological Situations in the Russian North.* St. Petersburg: State Polar Academy.

47 Georgievna, TY (2003). The account of losses at the oil recovery. The dissertation abstract of candidate's thesis in economy (St. Petersburg), p. 28.

48 *Ibid.*

49 Sergeevich, FA and K Vladimir (2009). Predicting Arctic climate: Knowledge gaps and uncertainties. In *Climate Change and Arctic Sustainable Development*, pp. 302–312. France: UNESCO.

50 Bedritskiv, AI (ed.) (2005). *The Strategic Forecasts of the Climate Changes of the Russian Federation for the Period till 2010–2015 and their Influence on Branches of Economy of Russia*, p. 28. Rosgidromet., *op. cit.*

51 *Ibid.*, p. 30.

52 *Ibid.*, p. 31.

53 Meehl, GA and TF Stocker. *Global Climate Projections*, Part 10, pp. 771, Fig. 10.13, pp. 773, Fig. 10.15.

54 Yevgenievich, FI, GZ Markovich *et al.* (2008). Sixty-year cyclic changes in the climate of the polar regions. *Materials of Glaciological Research*, 105, 158–165.

55 Petrovich, MA, SS Viktorovich and DD Mihailovich (2010). The modelling estimation of ice conditions in the Arctic basin in the end of XX and the XXI-st centuries. The thesis of reports of final session of Academic council AARI by results of the work in 2010, pp. 16–17.

56 *Ibid.*, p. 18.

SECTION II

WHAT ARE THE STAKES FOR THE LITTORAL STATES?

4

THE TRANSNATIONAL ARCTIC AND RUSSIA

Nadezhda Klimovna Kharlampyeva

4.1. INTRODUCTION

The discussions regarding the Arctic region in the early 21st century can be characterized within four policy-academic theses formulated by contemporary development and "open region" concepts. First, from the viewpoint of world political analysis, these postulates are:

(1) The Arctic is a region of peace and co-operation.
(2) The Arctic is a transportation route between Europe, Asia and North America.
(3) The Arctic region is one of mankind's bio-energy reserves.
(4) The Arctic region is one of Earth's ecosystems.

These four different approaches to the Arctic region shape Russians' understanding about its importance to Russia. They also constitute part of Russia's agenda for the 21st century, namely, promoting the formation of an Arctic region policy.

Second, the Arctic region is now included in "international relations", "world politics" and "international law" with its own new ideas, concepts and state strategies, i.e., assumptions and intentions concerning the potential of the Arctic region and its further reasonable use for the development and improvement of human well-being.

Third, the new situation around the Arctic region assumes interaction between state and non-state actors. Non-state actors, such as transnational companies and civil society, become main actors in national policy and international co-operation in their respective Arctic countries. At the moment, intergovernmental organizations are consolidating mechanisms for increased transnational action.

Thus, it is possible to say that the Arctic region today is an experimental platform for the introduction of international political decision-making innovation with a transnational character. National political and world political processes within the system of historical analysis involves state and non-state actors, interaction between them and with external players, requiring a transnational, global management approach. Russia's approach to the Arctic and its importance for Russian national interests can be discussed against this background. Russia currently regards the Arctic region as a particular sphere of its national interests and security.

4.2. RUSSIA'S PRINCIPAL VIEWS ON THE ARCTIC

For various reasons, the Arctic is strategically important for Russia. From the geopolitical point of view, Russia regards the Arctic as a sub-region of the North-Atlantic region and a transportation route between Europe, North America and Asia, where sea and air communications are controlled by navies. As an Arctic littoral state, Russia shares 22,600 km of its border with the Arctic, accounting for the longest section of the 38,700 km borders of all the littoral states. Sixty-four per cent of the Russian territories belong to its northern region, and this is equivalent to 50% of all the northern territories of the Arctic states where 85% of Russia's natural resources are extracted.

The geopolitical value of the Arctic region is described in two state documents — the Marine Doctrine of the Russian Federation which was adopted in 2001[1] and Bases of a State Policy for the Russian Federation in the Arctic Region for the Period till 2020 and Beyond, which was approved in 2008.[2] In the first document, the importance of the Arctic for Russia is identified within the context of its continental navigation, territorial claims, economic growth, wealth and military/security considerations. The covered issues are the Arctic's role in the maintenance of an open sea route for the Russian fleet to the Atlantic, the crucial role of the Russian Northern naval fleet for defending Russia against threats arising from the seas, the increasing value of the Northern Sea Route (NSR) for sustainable development of the Russian Federation, the importance of the Arctic's natural riches in the Russian exclusive economic region, and the Arctic's geographical/geological importance as part of the continental shelf of the Russian Federation.

The second document includes a discussion between the government and civil society on the protection of the Russian Federation border in the Russian Arctic region (not the High North as named earlier in the legal documents). The document also deals with maintaining a favourable operative navy in the Russian Arctic, including the maintenance of a necessary fighting potential for the Russian armed forces in this region.

Currently, the geopolitical situation in the Arctic region is perceived in Russia as "tending to increase rivalry for the control over the Arctic region, and regional militarization in such directions as strengthening of the Arctic powers' military presence in the region, strengthening multilateral military co-operation in the Arctic region, growth of military doctrines' quantity, the beginning of the armed forces' accelerated modernization by the Arctic powers, and the use of navies to protect economic interests in the region."[3] Within this context, two possible scenarios around the Arctic region development can be put forward. The first suggests "escalating the usage of power arguments by the leading powers, and also strengthening of militarization and external players coming into the region — Japan, the Peoples Republic of China, and Brazil. The second provides for the resolution of disputes concerning the Arctic region within the next 10–15 years as a result of actions by the international institutions (the United Nations, Barents Council, and Arctic Council [AC]) and bilateral diplomacy."[4]

4.3. COMPONENTS OF RUSSIA'S ARCTIC POLICY FORMATION

Russia's Arctic policy has four major components. On the one hand, they indicate the framework through which the Russians view the

Arctic. On the other hand, they have influenced Russian policy formation towards the Arctic. The following sections provide an account of the major aspects of these four components.

4.3.1. *The Arctic is a region of peace and cooperation*

As the predecessor of the current Russian state, the Soviet Union developed certain assumptions at the end of the 1980s for formulating its policy towards the Arctic, "The Arctic is a region of peace and cooperation". It, therefore, proposed two initiatives: The creation of a denuclearized world and peaceful co-existence. The "World Cruise"[5] was then organized as a demonstration of new intentions on the part of the representatives of the Soviet Union and Finland to cooperate in the Baltic Sea region.[6] As a result, a new stage in Russian–Finnish relations began and a Northern policy of the USSR with the support of Finland (being a country in the Arctic region) was formed.

Within this context, the delegation of the USSR in the 42nd session of the UN General Assembly announced the principles of a new common security system for co-operation between Russia and the West covering the military, economic, ecology and humanitarian spheres, applicable also to the Arctic.[7] This proposal for USSR-West co-operation was based on the Murmansk speech of General Secretary Mikhail Gorbachev about Northern co-operation.[8] As the two sides initiated co-operation in the North in the post-Soviet era, Evgeny Primakov, then the Minister of Foreign Affairs, stated in 1996: "The modern Northern co-operation model can be considered a contribution to the cause of Common European Security in the 21st century."[9] Against this background and according to the current Russian government-formulated policy towards the Arctic, Russia is presently engaging in the Arctic as "a region of peace and co-operation."[10]

Yet, it is important to understand the framework within which Russia operates with respect to the Arctic, as it affects the policy formulation towards that region, while interacting with other interested parties to seek resolutions to any conflicts that may arise. Despite the fact that the "battle over the Arctic" is talked about more often these days, the emerging conflict over the region can be solved through peaceful measures. Towards that end, one option could be using the inalienable rights of the states to individual or collective self-defence; the UN Security Council (UNSC) could be persuaded to set rules to be followed by all concerned parties.[11]

The peaceful settlement of disputes within the United Nations and the regional organizations about collective self-defence is being discussed once again. Today, Russia is finding a balance between the US and two powerful poles — Europe and China[12] for the formation of a multi-polar world system within its framework to resolve conflicts with respect to various issues, including conflicts over the Arctic.

However, state actors are not the only players in the Arctic. Due to this, the question is whether the regional and international organizations could play an essential part in a collective decision-making process concerning war and peace. Despite the fact that these are the main factors in international relations, determining whether war or peace should prevail, the role of the regional and international associations is also increasing in this region. Regional organizations such as the AC, the Barents Euro-Arctic Council (BEAC), the Council of Baltic States (CBSS) and the Nordic Council of Ministers (NCM) are actively involved in the process of collective decision-making.

Other regional organizations as well as international ones are becoming involved in certain issues of regional and, especially, international importance pertaining to the Arctic such as regulating the development of regional resources, the internationalization of the NSR and the Arctic region's security. These issues have become pressing questions for the United Nations, particularly UNSC, the European Union (EU) and NATO, some of whose member states have stakes in the Arctic. These entities are the main non-state actors to influence the decision-making process concerning the Arctic region.

It is interesting to look at the current geopolitical situation around the Arctic as summarized in Tables 4.1 and 4.2. The Arctic geopolitical chart has been created to demonstrate the significant engagement of the regional organizations in Arctic affairs. These are the AC, Barents Euro-Arctic Council, Council of Baltic States and Nordic

Table 4.1 The Arctic Geopolitical Chart.

		UNSC	NATO	EU	AC	Barents Council	Nordic Council	Baltic Council
1	Russia	x			x	x		x
2	US	x	x		x	o		o
3	Canada		x		x	o		
4	Norway		x		x	x	x	x
5	Denmark		x	x	x	x	x Greenland Faeroe	x
6	Sweden			x	x	x	x	x
7	Finland			x	x	x	x Aland	x
8	Iceland		x		x	x	x	x
9	Great Britain	x	x	x	o	o		o
10	France	x	x	x	o	o		o
11	Germany		x	x	o	o		x
12	Netherlands		x	x	o	o		o
13	Poland		x	x	o	o		x
14	Spain		x	x	o			o
15	Latvia		x	x				x
16	Lithuania		x	x				x
17	Estonia		x	x				x
18	Belarus							o
19	Italy		x	x		o		o
20	Rumania		x	x				o
21	Slovakia		x	x				o
22	Ukraine							o
23	Japan					o		
24	EU			x		x		x
25	India							
26	China	x						
27	Brazil							
28	South Korea							
29	Singapore							
	Total members and observers	5	17	16	8/6	7/9	5	12/10

Note: x — member-states; o — observers.

Source: Table created by Nadezhda K. Kharlampyeva based on the data available at the websites of the UNSC,[14] NATO,[15] EU,[16] AC,[17] Barents Council,[18] Nordic Council,[19] and Baltic Council.[20]

Table 4.2 The Arctic Regional Organizations and States' Collective Interests in Other Organizations.

No	International Organizations	States	Observers	UNSC		NATO		EU	
				Members	Observers	Members	Observers	Members	Observers
1	AC	8	6	2	2	4	6	3	6
2	BEAC	7	9	1	3	3	8	4	6
3	Nordic Council	5	—	—	—	3	—	3	—
4	Baltic Sea States Council	12	10	1	3	8	8	9	7

Source: Table created by Nadezhda K. Kharlampyeva based on the data available at the websites UNSC,[21] NATO,[22] EU,[23] AC,[24] Barents Euro-Arctic Council,[25] Nordic Council,[26] and Baltic Sea States Council.[27]

Council. The main actors of the Arctic geopolitical area are the "Arctic states" according to the AC Declaration.[13] Apart from the eight Arctic states (Iceland, Canada, Denmark, Norway, Russia, Sweden, USA, and Finland), the state actors include those countries, which are observers in the regional organizations, the non-Arctic countries, which are ready to become observers in the AC and a non-state actor, the European Commission wishing to become a permanent member of the Arctic regional organizations.

As Table 4.1 shows, "The Arctic Geopolitical area" consists of the EU and 28 states, of which five are UNSC members, 17 are NATO members, 16 are EU states, eight are Arctic states, seven are the states of the Barents region, five are Nordic Council countries, 12 are Baltic Sea countries, and six other states, namely Japan, China, India, Brazil, South Korea and Singapore.

Table 4.2 demonstrates the potential for co-operation among the UNSC, NATO and the EU within the Northern regional and international organizations given that all these entities include some Arctic countries or non-Arctic countries with stakes in the Arctic. Some of these countries, such as Russia and the USA have memberships in more than one entity.

For example, the AC includes two members of the UNSC, four members of NATO and three EU members. The BEAC consists of one UNSC member, two members of NATO and four EU members. The Nordic Council has three NATO members and three EU members. The Baltic Sea States Council includes one UNSC member, eight NATO members and nine EU members.

Potentially, these actors will form an Arctic common security regime in the future and an Arctic management structure. The collective character in promoting ideas and influencing the decision-making process of the organizations in which they are members demonstrates a possible trend for Arctic policy formation. EU members can easily utilize their rights in the shaping of the European Arctic policy. In order to do this, they can rely upon their membership in the Barents Euro-Arctic Council, the Baltic Sea States Council and the AC.

As the Arctic organization is dominated by Western states with a long history of collective co-operation through their various organizational and non-organizational ties, the prevailing situation is not highly favourable for Russia to play a major role in influencing the collective decision-making process in the Arctic. However, the UNSC permanent members (USA, Russia, Great Britain, France, and China) can play a major role in forming an Arctic policy by virtue of their status. In this context, Russia can play a significant role. Yet, given the limit of this potential, Russia may opt not to accept a model of regional co-operation for the Arctic.

In short, Russia's view of the Arctic Ocean as a major region of importance to its national interests opens several possibilities for creating a new foreign policy regime for the Arctic.

4.3.2. *The Russian Arctic is a transportation route between Europe, Asia and North America*

According to the Russian Maritime Doctrine up to the year 2020, the Russian geopolitical interests in the Arctic region are manifested in certain issues of vital interest to Russia. They include: free access of the Russian fleet to the Atlantic Ocean via the Arctic; ownership of the wealth within the Russian Federation's exclusive economic zone and continental shelf; the crucial role of the Russian Northern Fleet for the defence of the marine and ocean areas; and the increasing importance of the NSR for sustainable development of the Russian Federation.[28] The Russian Maritime Doctrine, which is applicable to the Arctic, establishes the basic principles of the national shipping policy, and includes the following: increasing the role of the Russian Federation as a Flag State in the seas; improving the national law on control over foreign vessels arriving in Russian ports; and preserving and improving the system of maritime personnel training. The doctrine provides a framework for expanding Russian sea activities based on certain principles, which, in Russia, have been named the "principles of continent — ocean — atmosphere — space"[29] and the "mechanisms of complex sea management".[30]

The principles of continent — ocean, atmosphere, space — are based on the following three objectives: (1) Russia's active participation in research, exploitation and the use of seas and their resources, (2) Russia's understanding of different activities at seas and oceans, and (3) Russia's acting in the "World Ocean" (i.e., all the oceans including the Arctic) as a littoral state and in accordance with the governing international laws. The respective structural model includes three spheres: (1) acquiring all the requirements for sea activities, such as infrastructure, navigational capabilities and security; (2) creating regional infrastructures for all Russian sea activities in all oceans and seas; (3) actively participating in multilateral and unilateral regimes concerning the "World Ocean".

As per the mechanisms of complex sea management, there are certain problems regarding the state management of sea activities which must be addressed. These include unclear areas of cooperation and competition between state entities: for example, the Russian navy and various maritime entities involved in economic activities. It has been suggested that the mechanisms of complex sea management[31] consider the use of the iterative (Agile Project Management — APM) method for the optimization of sea activities, i.e., the co-ordination of the military and economic activities to the greatest extent possible. The author suggests uniting all variants of Russian sea activity functions undertaken by different state military and non-military entities, including those by the navy, sea security services and all non-military fleets. Besides the issue of non-traditional functions of the Russian navy to protect Russian oil and gas resources, the NSR and the Sea Economic Activities (SEA) security in the Arctic region are now being discussed[32] with the Russian government.

Currently, Russia follows the Marine Activity Development Strategy (valid until 2030) while dealing with Arctic issues.[33] Within this context, the main issue is now the NSR management, which has both organizational and legal aspects to be addressed. The Transportation Ministry suggests returning to the Regulations for Navigation on the Northern Sea Route, which was confirmed by the USSR Ministry of Marine Sea Fleet in 1990 to manage the Northern Sea Route[34] and amending the regulations as required. Yet, the mentioned regulations cannot be used because the state-owned Arctic Administration has been liquidated and the state-owned companies operating the ice-breaker fleet and shipping companies have been privatized.

Another main direction of the Russian maritime policy affecting its Arctic activities is technology modernization. The United Shipbuilding Corporation (JSC "USC")[35] builds military and non-military ships and provides them with the required services during their life span. The Corporation includes 41 state and private companies and 450,000 employees.[36] There are also other ship-building and ship-repair enterprises consisting of 29 state and 12 private firms, where about 100,000 people are employed. Moreover, eight integrated state-private research institutes are being created to augment the ongoing process of the formation of the Krylovsky State Centre of Science located in Saint-Petersburg.[37] The question of "ensuring safety of maritime activities" is also being discussed within the government. Certain laws such as the Code of Maritime Trade (1999), the Law of the Internal, Territorial Waters and the Coastal Region of Russia (1998) and the Law of the Natural Monopolies (1995) will be changed with respect to the development of the NSR and the Russian Arctic Region.

4.3.3. *The Arctic region is one of mankind's bio-energy reserves*

This assumption has a certain meaning for Russia and especially its Northern section. Russia's Arctic has many natural resources which are very important to the Russian economy. They include the following:

— About 30 billion cubic meters of forest, 40% of the fresh water reserves and 50% of the marine products in the Northern part of Russia, a large number of Arctic animals of economic value, including more than one million reindeer, about 4,000 musks, 50,000 pink seagulls, 30,000 brand geese, 15,000 small swans (included the Red Book) and scores of walruses, Greenland seals, sea calves and beluga fish.
— 15,000 km^3 fresh water ice (which is the same amount of water found in Lake Baykal).
— 283,053.74 km^2 (4.1% of the entire Arctic zone) are Specially Protected Natural Sites (SPNS).

— 40% of Russia's gold, 60% of its oil, 60% of its gas, 60–90% of its coal, nickel, copper, antimony, cobalt, tin, tungsten, mercury, apatite and flogopolite, 90% of its chrome and manganese, and 97% of its platinum metal, ore diamonds and vermiculites.
— 30% of the world's gold, 20% of its platinum metal and nickel, 15% of its copper, 10% of its cobalt, 10% of its oil (without shelf), 30% of its gas, 50% of its apatite, 15% of its tin, 6–8% of its tungsten and mercury, and 35% of its niobium.
— 100% of Russia's diamonds, antimony, apatite concentrate, flogopolite, vermiculites, 98% of its platonoides, 95% of its gas, 90% of its nickel and cobalt, 60% of its copper and oil, and about 50% of its forest and fish products".[38]

4.3.4. *Russian's energy strategy: challenges or opportunities?*

A quantitative estimation of the Russian continental shelf's hydrocarbon resources was published for the first time in 1971. The preliminary data suggested the presence of total resources of oil and gas up to 100 billion tons of oil equivalent.[39] While this data proved to be accurate before the 1990s, after the disintegration of the USSR, the estimation work was stopped. Insignificant discoveries have since been made by Gazflotom, affiliated to the company "Gazprom" in the Pechora and Karsky Seas. Offshore deposits are not currently under development in the Russian Arctic shelf. State companies Gazprom and Rosneft have hesitated to suspend their operations in the Russian continental shelf (i.e., offshore oil and gas operations), preferring to focus instead on Russia's continental deposits. The other reasons are: high risk of offshore projects, lack of the required modern technologies domestically, high cost of developing the required infrastructure, financial problems, and also lack of encouraging legislation.

Russian scientists "doubt the implementation of the Ministry for Economic Development and Trade's plans, dated 2008, according to which Russia should extract the following from its shelf: In 2010 — 33 million tons of oil and 37 billion cubic meters of gas, by 2015 — 50 million tons of oil and 125 billion cubic meters of gas, by 2020 — 55 million tons of oil and 195 billion cubic meters of gas. It is considered that these plans may be fulfilled only in the case of creating a consortium with the participation of domestic and foreign investors with a predominant role for state companies."[40] Thus, it is obvious that in Russia there is a problem in organizing extraction of offshore oil and gas deposits. Russian companies are today involved in developing the continental oil and gas deposits, which are technologically easier. Russia already has the required technology and human resources and it is far less costly compared to the offshore deposits.

The melting of the Arctic ice and the possibility of having access to the region's oil and gas resources in the future have created global interest in these resources which in turn has created a demand for estimates of their size. Russia's estimates are now constantly being updated and the real extent of the regional and Russian fossil energy resources is being revealed. Non-Russian sources estimates are also being made.

The following tables were created on the basis of Russian and American research. It is necessary to point out that these estimates are tentative and subject to constant changes due to new data.

Table 4.3 provides comparative estimates of the resources in the Arctic Ocean as well as the Atlantic, Indian and Pacific Oceans.

Table 4.3 The Potential of Continental Shelf Oil and Gas Resources (billion tons of oil equivalents).[42]

Authors	Pacific Ocean	Indian Ocean	Atlantic Ocean	**Arctic Ocean**
Levchenko (1984)	9.00	35.30	44.32	—
Geodekyan (1988) with added estimates Gramberg and Suprunenko (1994)	16.04	57.40	74.36	**83.48**
Gramberg *et al.* (1998)	—	—	—	**110.0**

Source: Table created by Nadezhda K. Kharlampyeva based on a table created by Gramberg, IS and NV Laverov (2000). *The Arctic on the Third Century (resource potential and environmental problems)*, IS Gramberg and NV Laverov, p. 141. Saint-Petersburg: Science; See also Gramberg, IS, DA Dodin and VL Ivanov (1998). *Mineral — Raw base and environment of Russian Polar Circle — future of cities.* Proceeding of Second Conference in Syktyvkar, pp. 47–58.

The Arctic Ocean shelf's deposits are estimated at 83 to 110 billion tons of oil equivalent. As Table 4.3 shows, based on the first estimation, the Arctic Ocean shelf's reserves are considerably larger than those of the Pacific and Indian Oceans but not much larger than those of the Atlantic Ocean. The second total estimate of the Arctic Ocean's oil and gas reserves in oil equivalent (about 110 billion tons), includes more than 30 billion tons of potentially recoverable oil and gas.[41]

Russia has a large share of the Arctic's oil and gas resources. Table 4.4 shows the estimated oil–gas resources in 14 Russian Arctic zones, as provided by the US' Interior Geological Survey Department, namely the Circum-Arctic Resource Appraisal.[43] According to this report, "the total mean undiscovered conventional oil and gas resources of the Arctic are estimated to be approximately 90 billion

Table 4.4 Oil–Gas Reserves in Russia's Arctic Province (cubic meters).[45]

	Province Code	Province	Oil in million cubic meters	Gas in billion cubic meters
1	WSB	West Siberian Basin	582.83	18456.05
2	EBB	East Barents Basin	1177.50	8995.98
3	YK	Yenisey-Khatanga Basin	887.72	2831.85
4	LSS	Laptev Sea Shelf	495.32	922.46
5	BP	Barents Platform	326.79	742.74
6	TPB	Timan-Pechora Basin	265.06	256.73
7	LM	Lomonosov-Makarov	175.96	202.73
8	LA	Lena-Anabar Basin	304.12	59.68
9	NCWF	North Chukchi — Wrangel Foreland Basin	13.66	171.83
10	VLK	Vilkitskii Basin	30.412	162.65
11	NWLS	Northwest Laptev Sea Shelf	27.383	127.14
12	LV	Lena-Vilyii Basin	59.914	37.82
13	ZB	Zyranka Basin	7.60	42.66
14	ESS	East Siberian Sea Basin	3.14	17.53
		Total for the Russian Arctic	4357.409	33027.85

Source: Created by Nadezhda K. Kharlampyeva based on the data available in *Interior US Geological Survey* (2008). Circum-Arctic Resource Appraisal: Estimates of Discovered Oil and Gas North of the Arctic Circle. USGS Fact Sheets 2008–3049.

barrels of oil (equal to 14.3 billion cubic metres), 1,669 trillion cubic feet of natural gas (equal to 47.2 trillion cubic metres), and 44 billion barrels of natural gas liquids (equal to 6.9 billion cubic metres)."[44] As can be seen, the main part of the oil–gas reserves is concentrated in a coastal region of Russia's continental shelf (4375.409 million cubic meters of oil and 33027.85 billion cubic meters of gas). Thus, the main oil reserves are located in East Barents Basin (1177.50 million cubic meters) and Yenisey-Khatanga (887.72 million cubic meters) and the major reserves of gas are located in three provinces: Western Siberian Basin (18456.05 billion cubic meters), East Barents Basin (8995.98 billion cubic meters) and Yenisey-Khatanga Basin (2831.85 billion cubic meters).

Tables 4.5 and 4.6 contain the 1970s–1980s estimates of the Russian offshore and onshore oil and gas deposits. These estimates today are merely a reference point for new estimates. Apart from provinces mentioned in these tables, Barentsevo-Karsky Province, including the Shtokmanovsky, Leningrad and Rusanovsky deposits, and also continental deposits of East-Siberian Province are likely to be promising locations for oil and gas explorations in the Russian Arctic region. The Russian Arctic and Subarctic deposits are interesting to the small Russian companies which are interested in short-term investments.

Table 4.5 Potential and Available Gas Deposits in Barents–Kara Province (trillion cubic meters).[46]

Province	Potential resources	Available deposits
Stockman	3.2	2.54
Leningradskoe	3.0	1.05
Rusanovskoe	3.0	0.78
Total	9.2	4.37

Source: Table created by Nadezhda K. Kharlampyeva based on the table created by Dodin, DA (2005). *Sustainable Development of the Arctic (problems and perspectives)*, p. 58. Saint-Petersburg: Science.

Table 4.6 Oil and Gas Deposits in the Russian Arctic and Sub-Arctic.

Province	Oil (million tons), A + B + C,	Gas (trillion cubic meters)
Samotlorskoe	1,375	—
Priobskoe	746.2	0.1064
Komsomolskoe	812.9	0.0103
Kharvytinskoe	885.6	0.0310
Total	3,819.7	0.1477
		Gas deposits
Urengoyskoe		7.29
Yamburgskoe		4.78
Zapolyarnoe		3.53
Medvezhie		0.61
Bovanenkovskoe		4.92
Kharasaveyskoe		1.62
Yurubcheno-Tokhomskoe		0.70
Total		23.45

Source: Table created by Nadezhda K. Kharlampyeva based on the table created by Dodin, DA (2005). *Sustainable Development of the Arctic (problems and perspectives)*, p. 68. Saint-Petersburg: Science.[52]

The Russian oil and gas fields are situated mainly in the West-Siberian Province, which continues to Barents-Kara Province. Super-gigantic gas areas are found in the Bolshoy Urengoy and the North Urengoy (11 trillion cubic meters), the Yamburg-Kharvutinskoe (more than 6 trillion cubic meters), the Medvezhie (2 trillion cubic meters), and the Yamal-Nenets okrug (47 trillion cubic meters).

Table 4.6 identifies oil provinces with significant oil reserves such as Samotlorskoe (1375 million ton), Kharvytinskoe (885.6 million ton) and Komsomolskoe (812.9 million ton), and the large gas Province of Priobskoe (0.1064 trillion cubic meters). Gas supergiant provinces such as Urengoyskoe, Yamburgskoe, Zapolyarnoe and Medvezhie are considered to be the major supplying locations not only for Russia, but also the whole world.

According to the *Russia Energy Strategy 2030*, the hydrocarbon supplies in the Russian section of the Arctic should yield 96 million tons of oil and 320–330 billion cubic meters of gas production annually by 2020.[47] The most recent estimates of the Russian Arctic's oil and gas resources offer much larger figures than the previous ones: the Russian Arctic gas reserves are estimated at 164.2 trillion cubic meters, including those of the Russian continental shelf (63.8 trillion cubic meters) and the Russian deposits in the Siberian and other Northern provinces.[48] The hydrocarbon resources of Russia's continental shelf include 16.5 billion tons of oil and oil condensate.[49]

For the period ending 2030, the main areas of oil and gas production will be in the Russian continental provinces of Western-Siberian, Leno-Tungus, and Timano-Pechora. Therefore, the Russian oil and gas companies' policy is geared toward opening of new deposits and providing proof of the existence of new hydrocarbon reserves (larger than the existing ones) as a strategically important objective. Thus, "Rosneft" has opened a new deposit of oil in Moskovtseva (Khunty-Mansiysk Okrug) with estimated reserves of 33 million tons, 11.7 million tons of which are recoverable.[50] Work in this province will begin in 2012. Another Russian company, Arcticshelfneftgas, has released data on an increase in its Arctic oil supply for the State Commission on Energy Stocks. Suggesting the possibility of a further increase of 32% for its estimated reserve of 516.6 million tons, its recoverable supply indicates an increase of 75% for the total of 133.9 million tons, a larger reserve in comparison with earlier confirmed initial geological forecasts.[51]

The formation of a geo-economical area within the Russian Arctic region has two aspects. One is Russia's changing from playing a co-coordination-energy-trade role as it did in the Soviet era to acting as a liberal economic player in the world energy market.[53] During the planned economy years, the Soviet Union did not use liberal economy mechanisms. However, since the Soviet Union's fall in 1991, Russia has become a key player in global trade and this involves both state and private economic sector mechanisms. Thus, both state and private companies now operate in the Arctic oil and gas sector.

The other is the formation of a combined model for exploration and use of resources and territories.[54] The regional geographical infrastructure and socio-economical situations are different in the North compared to the other territories of Russia, demanding changes in co-management. For example, this reality has resulted in the quick development of the East-Siberian Oil–Gas Center, which is not typical of the development in other parts of Russia.

In practice, as shown by recent statistical data from Sakha (Yakutia) Republic "Using the Bowels of Earth's Plan in the East Siberian and Republic of Sakha (Yakutia) on 2005–2009", Siberian oil and gas platform has been built within three areas of the Russian Federation — Krasnoyarsk, Irkutsk and Sakha (Yakutia) regions. This has become possible due to joint interests of state and private oil–gas companies. Thus, the Kovyktansk and Angar-Lensk gas-condensate, Talakan gas-oil, Kujumbinsk oil–gas condensate, Jurubcheno-Tochom, Verchne-Chonsk and Chayanda gas-condensate-oil territories are called "South province of Siberian platform". The prospects of the province can be characterized by the 'East Siberian-Pacific Ocean (ESPO) oil/gas transportation system passing through this region.[55]

The Arctic projects accepted by the government for implementation demand changes in the national law. In particular, this concerns the raw mineral sector. Today, the law concerning investment in the raw material sector of Russia, including fossil energy, is not clear. The ongoing discussion over Russian federal law on "Bowels of the Earth" seeks to clarify the situation as it defines mineral deposits as a strategic category of resources requiring a balance between the foreign and domestic companies. The proposed law provides two options for setting the criteria for foreign involvement in developing Russian mineral resources: The first option is limiting foreign companies to deposits of oil over 150 million tons, gas over one billion cubic meters, gold over 700 tons, and copper over 10 million tons. The second option is decreasing the ceiling for oil to 70 million tons, for gas to 500 million cubic meters, for gold to 50 tons, and for copper to 500 thousand tons.[56]

4.3.5. *The Arctic region is one of Earth's ecosystems*

The political and academic postulate that the Arctic region is one of Earth's ecosystems has received considerable attention from the science-international co-operation policy nexus. Support for research activities geared to the accumulation and formation of new knowledge about the Arctic and the North[57] has promoted the institutionalization of international co-operation and the creation of the international area in the Arctic region. For the sake of clarity for non-Russian readers, it should be stated that there is a difference between what "Arctic" and "North" imply when used by the Russian government. "Arctic" includes sea management and also managing the economy of coastal regions, while "North" refers only to coastal region management. These definitions are now being discussed in the context of a new law, namely, the Law on the Russian Arctic Region. It has promoted the institutionalization of international cooperation and the creation of the international area in the Arctic region.

The Arctic region of the Russian Federation is declared to be a polygon for the development of fundamental and applied scientific research. Consequently, international co operation will be focused mostly on the accumulation of new knowledge relating to the Earth's global climate processes and stability of the Arctic ecosystem. As a main goal, we observe the introduction of institutional innovations and perspective models of production, such as special economic regions, clusters, business incubators, strengthening of export specialization of sub-Arctic territories, attraction of foreign investments and technologies, the organization of work in the economy of the Russian Arctic, and initiation of various inter-regional and international projects based on private-state partner principles.

With respect to the environment, the Arctic region's problems have been concentrated around the trans-boundary movement of hazardous wastes and the region's layer reduction. Working on these issues has been an example of international co-operation between scientists and politicians. As the practice of the international organizations' activities in the North and the Arctic shows, further co-operation will be focussed on making laws pertaining to issues of major concern for the Arctic. These include growing CO_2 emissions, sea environment protection, Arctic biodiversity, and protection of the Arctic's cultural and natural heritage.

In this case, the co-operation between politicians and scientists has been geared toward an interdependence to protect the environment and focus on balancing environmental concerns with social-economic activities. The benefit of the policy is to work together and take decisions on the effective measures of the sustainable management of development in the Arctic region. In order to find a workable compromise between the two mentioned interests, Russia supports the concept of sustainable development. Now, it is not clear how this idea should be realized at the legal and management levels. However, as far as the Russian scientists are concerned, discussing the absence of an agreement on the Arctic should not grant the countries, with stakes in the region, grounds "for irrational attitude to the Arctic resources, as according to the international law, any country, which explores natural resources of a region should take into the consideration the legitimate rights of other countries. In this case, there is a need to adopt standards for the preservation and development of the Arctic."[58] Thus, it is safe to say that Russia has no concrete program for maintaining ecological balance in the Arctic at this time.

This issue is currently being discussed within the limits of Russian government programs.

4.3.6. *The Arctic's international regime*

Within this context, debates on the Arctic's international regime are taking place in Russia. At least two major issues are of special importance. First, the socio-ecological and political problems of the Arctic region should be tackled in terms of an international regime and global management for the region. Second, relevant international laws should define the nature of the Arctic policy governing the Arctic's international regime. The Russian Arctic region with its large territory consisting of land and water areas falls within the regulating mechanisms of such an international regime. Having such a vision in mind, Russia is working to specify and establish its Arctic borders. It is also planning to reanimate check-points and security systems in the region, create models of private-state partnership in the regional economy and implement internationally-approved environmental projects. Thus,

Russia is prioritizing international requirements in conducting its policy towards the Arctic. Nevertheless, many of its activities in this region are understandably being carried out from the point of view of protection of the country's own national interests to meet international challenges.

Over the last few years, the international public has been widely discussing questions concerning the border demarcation of the Arctic Ocean and thus the issue of its littoral countries' continental shelf. Within this framework, the Arctic international regime is discussed in Russia from two standpoints: the historical lawmaking approach and the norms of the United Nations Convention on the Law of the Sea (UNCLOS) 1982.

The historical law-making approach stresses the necessity to reaffirm the historical rights of Russia in the Arctic region, giving special attention to the formation of a correct legal base concerning the Russian Arctic sector.[59] Along these lines, the formulation of the Russian Arctic policy needs to be guided mostly by the results of domestic scientific research taking into account international laws. Moreover, attention must be paid to protecting the Arctic environment through national legislative regulations. Within this environmental context, international contracts with Russian participation can be allowed.[60]

Concerning the Arctic passages, supporters of a historical law-making approach argue that "it is wrong to extend on international laws, applicable to other international passages, on the common bases."[61] They hold that "legal regimes of passages and the rights in them of the coastal states are differentiated depending on the type of specific passage."[62] As for the case of Russia and the Arctic, customary laws prevailing over the general rules of the UNCLOS 1982 are preferable.[63]

In order to preserve the Arctic's environment, proponents of the historical law-making approach support the application of Edict of Council of Soviet Union Presidium of 28 November 1984 about Far North nature protection in the Russian Arctic sector, taking into account the international environment law.[64]

The Edict covers certain areas, i.e., "islands of the Arctic Ocean and its seas, the islands of Beringova and Okhotsky seas, other territories of the Soviet Union under the Ministerial Council, and the seas (including areas around the islands, belonging to the Soviet Union), if they influence the ecological well-being of the Soviet Union Northern part."[65]

The norms of the UNCLOS 1982 focus on the delimitation principles of the Arctic Ocean seabed and subsoil of the submarine areas between the Arctic neighbouring states based on the norms of the UNCLOS 1982 to determine Russia's Arctic territory.[66] The principle of "baselines" according to item 4(a) of Article 76 of the UNCLOS 1982 or principle of "equidistant of media line" according to item 2 of Article 2 of the Convention on the Continental Shelf 1958 take into account "special circumstances" regarding the issue of delimitation, and to a lesser extent, the principle of "Common heritage of mankind" as reflected in Article 136 of the UNCLOS 1982. The authors of the UNCLOS 1982 confirmed that Russia, as the coastal state, has the right to accept certain stricter measures than those accepted at the international level within its exclusive economic zone (EEZ). These are (a) to pass a domestic Law about the NSR, (b) establish a permission requirement for the passage through the NSR of foreign naval vessels and other foreign vessels having nuclear reactors or transporting nuclear materials and other dangerous substances to the maritime environment, (c) to forbid foreign military aircraft flights over the NSR, which transport dangerous materials, (d) to require permission for air photography, hydro engineering and other research carried out in the NSR, including the area leading to its EEZ, territorial sea and internal sea waters and their directly adjoining coast, and (e) to declare the NSR a special, vulnerable area following the example of the Baltic countries according to item 6 of 211 Article of the UNCLOS 1982.[67]

4.3.7. *Global governance of the Arctic*

Russia has been interested in the concept of global governance over the Arctic. In this regard, certain ideas have been promoted by Russia and other concerned countries. They include expanding the membership by inviting the EU countries into the AC, and the creation of "ocean management" by scientific units of the United Nations and the "Arctic Five", which includes Russia and some other Arctic states. This concept has become meaningful in the continuous process of interaction between Russia and the other Arctic states, involving both private and public actors. This interaction improves the quality and effectiveness of the principles of collective management, the distribution of duties and responsibilities, the degree of correspondence between domestic and foreign policies of states, and the interaction of private companies and social movements in the institutional, functional and structural context.

During the first stage of Arctic co-operation among international organizations since the 1980s, initiatives concerning the state of the environment and life of the indigenous peoples have been proposed for discussion in future Arctic Agendas. Active international cooperation in the North has coincided with the energy discourse and climate change in the Arctic.[68] The resulting general concern has led to the adoption of the "format of four councils" based on the outcome of the Moscow Meeting of Committee Heads of the Northern Regional Organizations held in 2006.[69] Hence, four supranational bodies, namely the AC, CBSS, BEAC and NCM, agreed to initiate a project for the purpose of efficient work in the future in the Arctic. Table 4.7 shows the main directions and contact points for international co-operation and the most applicable regional differentiation concerning various issues.

Table 4.7 The Principle of the "Format of Four Councils".

	AC	BEAC	NCM	CBSS
Climate change	+	+		
Energy	+	+	+	
Local peoples	+	+	+	
Frontier customs co-operation		+	+	+
Democracy, civil society			+	+
Emergency prevention	+	+		+
Exchange of experience between secretariats			+	+

Source: Created by Nadezhda K. Kharlampyeva based on The sixth annual meeting of CEO regional international organizations of the North (2006). Moscow, Russia, August 18, *Arctic Idea*, September. pp. 14–15.

Thus, the Nordic States are involved in implementing the frontier and customs co-operation project. First, it is linked with business, trade and investment activities. The issues of democracy and establishing civil society are developed by the Scandinavian nations and the Baltic and North-West regions of the Russian Federation. Emergency prevention measures are urgent for sea coastal areas of both the Arctic Ocean and the Baltic Sea. However, the CBSS is not aimed at solving such problems as climate change, energy and indigenous people. The AC is not engaged in issues such as frontier and customs co-operation, democracy and civil society, and exchange of experience between secretariats. Its priorities include climate change, local people, energy and preventing emergency situations. It is noteworthy that at the ongoing first stage of international co-operation, the activities of the regional and international organizations have followed eight main directions: protection of the environment and provision of security, sustainable development, healthcare and social issues, cultural dimensions, youth linkages, information and communication technology development, cooperation of parliamentarians, and research/scientific and educational projects.[70]

4.4. CONCLUSION

For Russia, the Arctic region is of crucial importance for several reasons. It is of military/security importance as it is the strategic part of the North Atlantic region and provides a free exit for the Russian fleet to the Atlantic. It is also a region for deploying the Russian Northern Fleet necessary for the maintenance of Russia's national security. The Arctic is also important for its potential role in intercontinental trade and navigation geared to the Northern Sea NSR. This route could potentially rival the existing Southern Sea Route via the Straits of Malacca and the Suez Canal. The Arctic is also crucial for its vast energy resources (oil and gas) which could have a major impact on global energy markets. The Arctic is also very significant for its vast mineral resources. In terms of the environment, the continued melting of the ice could have major negative long-term consequences not only for Russia and other Arctic littoral states, but for all other countries.

Today, the Arctic's attraction for Russia, like other littoral states, is geared especially to its economic and energy potential, namely oil/gas and mineral extraction and navigation. Its economic appeal places new tasks before Russia and other Arctic states: specification of sea borders, restoration of marine fleets, development of the Northern economy and building the required infrastructure. However, major activities in these areas are subject to the continued melting of the Arctic, making major energy and mineral extractions possible as well as reliable use of the NSR. Having said this, the ongoing climate change, which probably has a cyclic character of warming and cooling on coastal territories of the Arctic region, is not a guarantee for prospective year-round navigation or safe extraction of hydrocarbon raw materials. Thus, it can be assumed that interest in the natural resources of the Russian Arctic region will increase in the near future provided the melting process continues for a long period of time.

BIBLIOGRAPHY

About Environmental Protection (Moscow, Russia) December 20, 2001. This Federal Law was adopted in State Council in December 20, 2001, in Federal Council December 6, 2001, entered January 1, 2005.

Agreement between the Government of Canada and the Government of the Russian Federation on cooperation in the Arctic and the North E100317-CTS 1992 No. 18. http://www.treaty-accord.gc.ca/texttexte.asp?id=100317 [accessed 21/11/2011].

Agreement on cooperation on aeronautical and maritime search and rescue in the Arctic. Arctic Council. Nuuk.

Arcticshelfneftgas Increase Deposit. *InterEnergo Portal.ru.* http://ieport.ru/9195-rosneft-otkryla-v-xmao-novoe-neftyanoe.html

Basic of the State Policy of the Russian Federation in the Arctic for the period till 2020 and for further perspective (Moscow, Russia), Adopted by the President of the Russian Federation, D Medvedev, 18 September 2008. Promulgated: 30 March 2009, publication of the official governmental newspaper "Rossiyskaya Gazeta".

Bynic, IV. *Substantiation of the Russian Federation Rights in Arctic Region: The Review of the Legal Doctrine.* Russia: Timofeev, Vahrenwald& Partners LLP.

CBSS Observer States. *Council of the Baltic Sea States.* http://www.cbss.org/Cooperation/cbss-observer-states

Countries and Territories. *Nordic Council.* http://www.norden.org/en/about-nordic-co-operation/countries-and-territories

Dodin, DA (2005). *Sustainable Development of the Arctic* (problems and perspectives). Saint-Petersburg: Science.

Declaration on the occasion of the Seventh Ministerial Meeting on the Arctic Council (2011). *Arctic Council.*, 12 May, Nuuk, Greenland. http://arctic-council.org/filearchive/Arctic SAR Agreement EN FINAL for signature 21-Apr-2011.pdf

Gramberg, IS (ed.) (2000). *The Arctic in the Third Century* (*Resource Potential and Environmental Problems*). Saint-Petersburg: Science.

Gramberg, IS, DA Dodin and VL Ivanov (1998). Mineral — raw base and environment of Russian polar circle — future of cities. *Proceedings of II Conference in Syktyvkar.*

Graziani, T (2011). Geopolitical tensions and the multipolar system: The US versus Eurasia global research. *Global Research*, May 13. http://theglobalrealm.com/2011/05/13/geopolitical-tensionsand-the-multipolar-system-the-us-versus-eurasia/

Kalamkaryan, RA (2008). *Codification of International Law and Contemporary Agenda*, p. 248. Moscow: Nauka.

Kaminsky, VD (2011). What is the Russian shelf preparing for US? *Oil of Russia*, 9, 3.

Kimmelman, SA and LI Koshkin (2011). Problems of state primary management. *Management in Russia and Abroad*, 1, 17–25.

Kolodkin, AL, VN Gutsulyak and YV Bobrova (2007). World Ocean. In *International Legal Law: The Basic Problems.* Moscow: Statute.

Konyshev, VN and AA Sergunin (2010). Arctic on the Line of Geopolitical Interests. *World Economy and International Relations*, 9, 43–53.

Korzun, VA (2005). *Russian Interests in the World Ocean.* Moscow: Nauka.

Levintov, AE (2007). Model of exploration and using territory and resources. In *The North: Problem of Periphery Territories.* Syktyvkar, Russia.

Lukyanov, F (2010). Russian Dilemmas in a multipolar world. *Journal of International Affairs*, 63(2), (Spring/Summer 2010), 19–32.

Marine Doctrine of the Russian Federation (2002). Ministry of Defense and Marine Navy, 27 July. St. Petersburg: Press GUNIO.

Markushina, NY and NK Kharlampyeva (2008). *North-West Federal Okrug: External Affairs.* St-Petersburg: Press St-Petersburg University.

Makeev, BN (2008). Several methodological problems of sea activity. *World Economy and International Relations*, 8, 37–48.

Maleev, YN and VA Sokirkin (2000). Arctic passages as an aspect of international passages Sui Generis. *Moscow International Law Magazine*, 4, 223.

Member States. *Arctic Council.* http://arctic-council.org/section/member states

Members States of the EU. Gateway to the European Union. http://europa.eu/about-eu/countries/index en.htm

Members of the Barents Euro-Arctic Council. *The Barents Euro-Arctic Council.* http://www.beac.st/in English/Barents Euro-Arctic Council/Introduction/Members.iw3

Members, Membership in 2011. *UN Security Council.* http://www.un.org/sc/members.asp

Minaeva, IY (2004). Ecological security of the Arctic region: International law aspects. *Siberian Law issues*, 1. http://www.law.edu.ru/doc/document.asp?docID=1160070&subID=100047090,100047088

NATO Member Countries. http://www.nato.int/cps/en/SID-EAF18491-6768DB00/natolive/nato countries.htm

Non-Arctic States. *Arctic Council.* http://arctic-council.org/section/observers non arctic states

Observers of the Barents Euro-Arctic Council. *The Barents Euro-Arctic Council.* http://www.beac.st/in English/Barents Euro-Arctic Council/Introduction/Observers.iw3

Patrushev, NP. Need Modernization for Ship-building Sphere (10 April 2010). Interview Secretary of Russian Security Council, *Izvestiya*, 104.

Putin VV (2010). Need Navy Fleet Develop program. The Address in meet ing JSC "Admiralteyskie Verfi". (St. Petersburg, Russia), 18 December 2009, *Russian Military survey*, 1.

Primakov, EM (1996). The address in session of Barents Euro-Arctic council. Petrozavodsk, Russia.

Regulations for Navigation in the Seaways of the Northern Sea Route (1990), 14 September. http://www.morflot.ru

Rosneft Opens New Deposit in Khanty-Mansijsk. *InterEnergo Portal.ru* http://ieport.ru/9195-rosneft-open–in -khanty-new-oil.html

Russian Marine Activity Development Strategy till 2030 order of the Government of the Russian Federation (2010). 8 December. http://www.rosmorport.ru/media/File/filials/Central%20Office/2205-r.pdf

Sinetsky, VP (2008). Sea activity: Theoretical aspects. *World Economy and International Relations*, 8, 49–56.

Site of the Krylovsky State Centre of Science. http://www.ksri.ru/rus/cnt/ins.htm

Gert, A (ed.) (2010). *Using of the Earth's Bowels in East-Siberian and Sakha (Yakutia)*. 1 January (Novosibirsk), p. 3.

The address of the head of Soviet Union delegation in UN (25 September 1987). *Pravda* , 268 (252555), 4.

The Address Yuriy Trutnev. Ministry of Russian Federation Ministry of Natural resources. http://www.mnr.gov.ru/press-service/publications/detail.php?ID=11089

The Address of Michail S. Gorbachev in Murmansk (2 October 1987). *Pravda*, 275 (25262), 1–3.

The Council. *Council of the Baltic Sea States.* http://www.cbss.org/CBSS The-Council/the-council

The 6th Annual Meeting of Chairs of CEO Committees of the North Regional International Organizations (2006). Moscow, Russia, 18 August. *Arctic Idea*, September. pp. 14–15.

To participants of Soviet-Finland Baltic Sea Cruise (1 September 1987). *Pravda*, 244 (252311), 1.

United Shipbuilding Corporation. *Charter of the Joint Stock Company.* (JSC "USC"). http://www.oaoosk.ru/docs/ustav change/ustav.pdf

USGS. *Circum-Arctic Resource Apprasial: Estimates of Discovered Oil and Gas North of the Arctic Circle.* Department of the Interior of US Geological Survey, USGS Fact Sheets 2008–3049.

Voronov, KV (2010). Arctic horizons of Russian strategy: Modern dynamic. *World Economy and International Relations*, 9, 64.

Vylegzhanin, AN (2010). Formation of global legal space in the XXI century. *International Processes*, 2(8), 23.

Zhiznin, SZ (2005). Energy Diplomacy: Economy. *Policy and Practice* (Moscow), p. 78.

ENDNOTES

1 *Marine Doctrine of the Russian Federation* (2002). Ministry of Defense and Marine Navy, 27 July. St. Petersburg: Press GUNIO.

2 *Basic of the State Policy of the Russian Federation in the Arctic for the period till 2020 and for Further Perspective* (Moscow, Russia), Adopted by the President of the Russian Federation, D Medvedev, September 18, 2008, Promulgated: March 30, 2009, publication of the official governmental newspaper "Rossiyskaya Gazeta".

3 Konyshev, VN and AA Sergunin (2010). Arctic on the line of geopolitical interests. *World Economy and International Relations*, 9, 43–53.

4 Voronov, KV (2010). Arctic horizons of Russian strategy: Modern dynamic. *World Economy and International Relations*, 9, 64.

5 To participants of Soviet-Finland Baltic Sea Cruise (1 September 1987). Pravda 244 (252311), pp. 1. World Cruise is the name of a Russian–Finnish political action. Thus, Finland's and Russia's civil societies organized a Baltic Sea cruise as an example of close co-operation after "cold war" between West and East.

6 *Ibid.*

7 The address of the head of Soviet Union delegation in UN (25 September 1987). *Pravda* 268 (252555), p. 4.

8 The address of Michail S. Gorbachev in Murmansk (2 October 1987). *Pravda* 275 (25262), pp. 1–3.

9 Primakov, EM (1996). The address in session of Barents Euro-Arctic council. *Petrozavodsk*, Russia.

10 *Basic of the State Policy of the Russian Federation in the Arctic for the Period till 2020 and for Further Perspective* (Moscow, Russia), Adopted by the President of the Russian Federation, D Medvedev, 18 September 2008, Promulgated: 30 March 2009, publication of the official governmental newspaper "Rossiyskaya Gazeta", *op.cit.*

11 Kalamkaryan, RA (2008). *Codification of International Law and Contemporary Agenda*, pp. 248. Moscow: Nauka.

12 Lukyanov, F (2010). Russian dilemmas in a multipolar world. *Journal of International Affairs*, 63(2), 19–32.

13 *Arctic Council Declaration*, signed by the representatives of the Arctic States in Ottawa on the 19th of September 1996.

14 Members, Membership in 2011. UN Security Council. http://www.un.org/sc/members.asp

15 Member Countries. *NATO.* http://www.nato.int/cps/en/SID-EAF18491-6768DB00/natolive/nato countries.htm

16 Members States of the EU. *Gateway to the European Union.* http://europa.eu/about-eu/countries/index en.htm

17 Member States. *Arctic Council.* http://arctic-council.org/section/ member states

18 Members of the Barents Euro-Arctic Council. The Barents Euro-Arctic Council. http://www.beac.st/in English/Barents Euro-Arctic Council/Introduction/Members.iw3

19 Countries and Territories. *Nordic Council.* http://www.norden.org/en/about-nordic-co-operation/countries-and-territories

20 The Council. *Council of the Baltic Sea States.* http://www.cbss.org/CBSS-The-Council/the-council

21 Members, Membership in 2011. *UN Security Council, op. cit.*

22 Member Countries. *NATO. op. cit.*

23 Members States of the EU. Gateway to the European Union. *op. cit.*

24 Non-Arctic States. Arctic Council. http://arctic-council.org/section/observers non arctic states

25 Observers of the Barents Euro-Arctic Council. The Barents Euro-Arctic Council. http://www.beac.st/in English/Barents Euro-Arctic Council/Introduction/Observers.iw3

26 Countries and Territories. *Nordic Council.* http://www.norden.org/en/about-nordic-co-operation/countries-and-territories

27 CBSS Observer States. *Council of the Baltic Sea States.* http://www.cbss.org/Cooperation/cbss-observer-states

28 *Marine Doctrine of the Russian Federation* (2002). Ministry of Defense and Marine Navy, 27 July. St. Petersburg: Press GUNIO.

29 Sinetsky, VP (2008). Sea activity: Theoretical aspects. *World Economy and International Relations*, 8, 49–56.

30 Makeev, BN (2008). Several methodological problems of sea activity. *World Economy and International Relations*, 8, 37–48.
*Definition "World Ocean" includes South Ocean.

31 Makeev, BN (2008). Several methodological problems of sea activity. *World Economy and International Relations*, 8, 37–48. *op. cit.*

32 Korzun, VA (2005). *Russian Interests in the World Ocean.* Moscow: Nauka.

33 *Russian Marine Activity Development Strategy till 2030 Order of the Government of the Russian Federation* (2010), 8 December. http://www.rosmorport.ru/media/File/filials/Central%20Office/2205-r.pdf

34 *Regulations for Navigation on the Seaways of the Northern Sea Route* (1990), 14 September. http:// www.morflot.ru

35 Charter of the Joint Stock Company "United Shipbuilding Corporation" (JSC "USC"). http://www.oaoosk.ru/docs/ustav change/ustav.pdf

36 Putin, VV (2010). Need Navy Fleet Develop program. The Address in meeting JSC "Admiralteyskie Verfi" (St. Petersburg, Russia), 18 December 2009, *Russian Military Survey*, 1, 2010.

37 Interview Secretary of Russian Security Council Nikolay P.Patrushev: Need modernization for ship-building sphere (10 April 2010). *Izvestiya*, 104; See also site of the Krylovsky State Centre of Science. http://www.ksri.ru/rus/cnt/ins.htm

38 Dodin, DA (2005). *Sustainable Development of the Arctic* (*Problems and Perspectives*), p. 58. St. Petersburg: Science.

39 Kaminsky, VD (2011). What is the Russian shelf preparing for US? *Oil of Russia*, 9, p. 3.

40 *Ibid.*

[41] Gramberg, IS (ed.) (2000). *The Arctic on the Third Century* (*Resource Potential and Environmental Problems*), p. 139. Saint-Petersburg: Science; See also Gramberg, IS, DA Dodin, VL Ivanov (1998). Mineral — Raw base and environment of Russian Polar Circle — future of cities. Proceeding of II Conference in Syktyvkar, 47–58.

[42] *Ibid.*

[43] USGS. *Circum-Arctic Resource Apprasial: Estimates of Discovered Oil and Gas North of the Arctic Circle*, Department of the Interior of US Geological Survey, USGS Fact Sheets 2008–3049.

[44] *Ibid.*

[45] *Ibid.*

[46] Table created by Nadezhda K. Kharlampyeva based on the a table created by Dodin, DA (2005). *Sustainable Development of the Arctic* (*Problems and Perspectives*), p. 58. St. Petersburg: Science.

[47] *The Russia Energy Strategy 2030*, adopted by Russian Federation Government, 13.11.2009.

[48] *Ibid.* With deposit in continent (in the Siberian and other provinces).

[49] *Ibid.*

[50] Rosneft opens new deposit in Khanty-Mansijsk. *InterEnergo Portal.ru* http://ieport.ru/9195-rosneft-open –in - khanty-new-oil.html

[51] Arcticshelfneftgas Increase Deposit. *InterEnergo Portal.ru* http://ieport.ru/9195-rosneft-otkryla-v-xmao-novoe-neftyanoe.html

[52] *Ibid*, p. 68.

[53] Zhiznin, SZ (2005). *Energy Diplomacy: Economy, Policy and Practice*, p. 78.

[54] Levintov, AE (2007). Model of exploration and using territory and resources. In *The North: Problem of Periphery Territories*, pp. 20–35. Syktyvkar, Russia.

[55] Gert, A (2010). *Using of the Earth's Bowels in East-Siberian and Sakha* (*Yakutia*), 1 January, Novosibirsk, p. 3.

[56] *The Address Yuriy Trutnev.* Ministry of Russian Federation Ministry of Natural resources. http://www.mnr.gov.ru/press-service/publications/detail.php?ID=11089

[57] As terminologies "Arctic" and "North" are used in two different geopolitical contexts in Russia, i.e., "Arctic" in the international context and "North" or "High North" in the national context (covering Russian Arctic region).

[58] Minaeva, IY (2004). Ecological security of the Arctic region: International law aspects. *Siberian Law issues*, 1. http://www.law.edu.ru/doc/document.asp?docID=1160070&subID=100047090,100047088

59 Bynic, IV. *Substantiation of the Russian Federation Rights in Arctic Region: The Review of the Legal Doctrine.* Timofeev, Vahrenwald& Partners LLP.

60 *Ibid.*

61 Maleev, YN and VA Sokirkin (2000). Arctic passages as an aspect of international passages Sui Generis. *Moscow International Law Magazine*, 4, 223.

62 *Ibid*, p. 225.

63 *Ibid.*

64 Vylegzhanin, AN (2010). Formation of Global Legal Space in the XXI Century. *International processes*, 2(8), p. 23.

65 *Ibid.*

66 United Nations Convention on the Law of the Sea of 10 December 1982 as "Constitution of Seas" is considered for the Arctic Ocean as one of regulation mechanisms according to Article 234. "Ice-covered areas: Coastal States have the right to adopt and enforce nondiscriminatory laws and regulations for the prevention, reduction and control of marine pollution from vessels in ice-covered areas within the limits of the exclusive economic zone, where particularly severe climatic conditions and the presence of ice covering such areas for most of the year create obstructions or exceptional hazards to navigation, and pollution of the marine environment could cause major harm to or irreversible disturbance of the ecological balance. Such laws and regulations shall have due regard to navigation and the protection and preservation of the marine environment based on the best available scientific evidence."

67 Kolodkin, AL, VN Gutsulyak, YV Bobrova (2007). *World Ocean. International Legal Law. The Basic Problems*, pp. 274–275. Moscow: Statute.

68 Markushina, NY, NK Kharlampyeva (2008). North-West Federal Okrug: External Affairs. St. Petersburg: St. Petersburg University Press.

69 The 6th Annual Meeting of Chairs of CEO Committees of the North Regional International Organizations (2006), 18 August, Arctic Idea, September. pp. 14–15.

70 *Ibid*, pp. 14–15.

5

NORWAY AND RUSSIA: NEIGHBOURS WITH STRONG INTERESTS IN THE ARCTIC

Arild Moe

5.1. INTRODUCTION[1]

Both Norway and Russia are Arctic states as well as big petroleum producers.[2] In both states, the Arctic is considered very important for the further development of their offshore petroleum sectors. This chapter presents and discusses main developments with respect to the Norwegian and Russian offshore petroleum policy in the Arctic and what has characterized the relationship between the two countries in this sphere: competition, conflict, or co-operation. Questions have been raised about the stability of the jurisdictional arrangements in the Arctic. This chapter, therefore, discusses major and recent developments with regard to jurisdiction and delimitation of the Arctic continental shelf, focusing on Russia and Norway. It seeks to answer certain questions of major importance for both the regional peace and security, and the development of the Arctic's mineral and energy resources, particularly oil and gas, namely, how can the recent delimitation agreement between Norway and Russia in the Barents Sea be explained? And, what is the outlook for conflict surrounding the Russian claim for an extended shelf in the Arctic Ocean?

5.2. OFFSHORE PETROLEUM ACTIVITIES IN THE HIGH NORTH

5.2.1. *Norway*

Most of Norway's oil production takes place in the North Sea, where oil output is now falling. Over the past decade, the biggest contribution to new resources has come from the Norwegian Sea, outside the "middle part" of Norway. For future production, the focus will increasingly be on the Northern part of the continental shelf, the Barents Sea (see Fig. 5.1). When output from existing gas fields in the South starts to decline, a construction of pipelines up to the Barents Sea is conceivable. Thus Barents Sea gas could fill up free capacity in the extensive gas pipeline network, which is already connecting the Norwegian continental shelf in the North Sea with terminals on the European continent and the UK.

Until the early 1980s, Norway was reluctant to start exploration activities in the High North for two main reasons: the new petroleum nation had plenty to do further south where vast oil and gas resources had been discovered and its relations with the Soviet Union played a part. At that time, the USSR signalled that it did not want to see international oil companies in the Barents Sea, and the Norwegian development plans took this into consideration by prescribing dominance by Norwegian companies. However, this line of policy was abandoned in the early 1980s and the major oil companies were hence invited to apply for exploration licenses.[3] In the mid-1980s, there was widespread optimism about the resource potential of the Norwegian Barents Sea; thus a surge in activity was expected. However, the results of the exploration efforts were not very impressive and at times the oil companies were ready to give up on the region. Since production in the area is expected to be relatively expensive, interest in the area has also been reduced over a period of time with low oil prices.

In addition to the oil industry, the Norwegian government has been uncertain about the path forward. Possible conflicts with fisheries' interests and environmental issues have loomed large in the internal debate, whereas regional development has been an argument

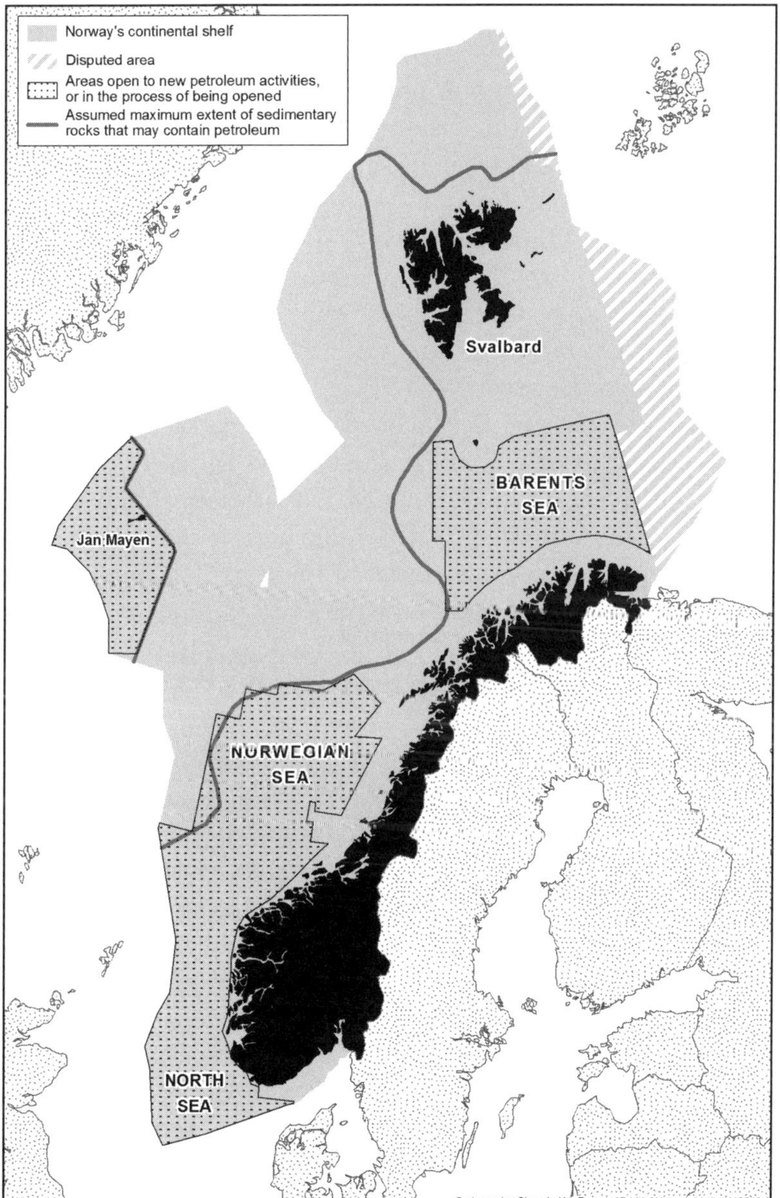

Fig. 5.1. The Norwegian Continental Shelf.

Source: Created by Claes Lykke Ragner, Fridtjof Nansen Institute based on illustration in Petroleum resources on the Norwegian continental shelf 2009, Norwegian Petroleum Directorate, 2009.

for increased activity. By 2001, the heated debate prompted the Norwegian government to temporarily ban all new exploration licensing in the area. The ban was partly lifted in 2003, and a series of blocks were licensed in three rounds from 2003 to 2009. Clearly, the government now wanted to increase activity, balancing environmental and economic considerations as discussed. The latest decisions indicate that the authorities perceive a sufficient interest from the oil industry to go forward with exploration in the Barents Sea.

At the same time, there is a growing consensus that large discoveries are not very likely to happen. Several, but small, finds are seen as probable. However, making such finds commercially viable will require coordinated development of infrastructure as well as resources. Exploration activity in the Barents Sea has not been very intensive, however, with only some 90 exploration wells drilled. The size of expected undiscovered recoverable petroleum resources in the (earlier) undisputed Norwegian part of the Barents Sea is 910 billion (bn) standard cubic meters of oil equivalent (scmoe) or 764 million tons of oil equivalent (mtoe), slightly more gas than oil. The estimated undiscovered volumes in the North Sea and Norwegian Sea are about the same, but those parts of the Norwegian continental shelf have much larger proven reserves. The degree of uncertainty in the Barents Sea is high since there has been no drilling in large areas.

To date, several minor discoveries have been made, but only one field, the *Snøhvit* ("Snow White") gas field, discovered in 1984, has been developed. The second substantial discovery, the *Goliat* oil field, was made in 2000 by Agip. An oil discovery by Statoil — *Skrugard* — which was publicized in 2011 is also very promising and is already attracting new industry interest for the region.[4]

The Snøhvit field came on stream in the fall of 2007 as the first large liquefied natural gas (LNG) project in Europe. It is located 140 km from shore northwest of Hammerfest in Norway's northernmost county, Finmark. The development of the project was stopped several times as the market outlook was not deemed promising. Only after 2000, with positive expectations in the US market and also after tax concessions, did it go ahead. It was still a difficult birth involving commercial uncertainty and cost overruns; it experienced serious technical problems right at the beginning.[5] The field, with recoverable reserves of 161 bn standard cubic meters (scm) of gas and 18 million scm of gas condensate, developed with sub-sea installations at water depths of about 300 meters, is being considered a pioneer project. The gas is piped to shore where a processing plant has been built. From there, the gas is shipped as LNG in special carriers to market.[6] The field is slated to produce approximately 6 bn scm annually. Of this, 2.6 bn scm has been contracted by Statoil for the US market and 1.6 bn scm for Spain, whereas 1.7 bn scm will be sold by the other main partners, Total and GDF Suez.

In short, the Norwegian exploration effort and supporting activities in the Barents Sea are characterized as being eager but also ready to stop, and at the same time, the government would not hesitate to impose strong restrictions.

5.2.2. *Russia*

The Soviet Union started seismic surveying in the Barents Sea in the 1970s. In the 1980s, a systematic exploration effort took place revealing gas fields in the "super-giant" category in the North-Western part of the Russian Barents Sea. In the southeastern part, a number of promising structures were identified and some smaller oil fields discovered (see Fig. 5.2). In the 1990s, exploration activity fell drastically, for financial and organizational reasons. Altogether only about 60 wells have been drilled in the whole Russian Barents Sea, but the discovery rate has been very high. The outlook for a further concentrated exploration effort as well as industrial development on the Russian side of the Barents Sea seems better than on the Norwegian side. All of the 11 discoveries in the Russian Barents Sea are now designated as fields, meaning that their resources are expected to be recoverable. They consist of four oil fields, one oil and condensate field, three gas and condensate fields, and three gas fields.[7] According to Russian estimates, there are some 3,700 mtoe of recoverable resources in these fields, which have been examined in some detail.[8] This is more than the remaining reserves[9] on the entire Norwegian continental shelf. Natural gas is predominant in the Russian fields,

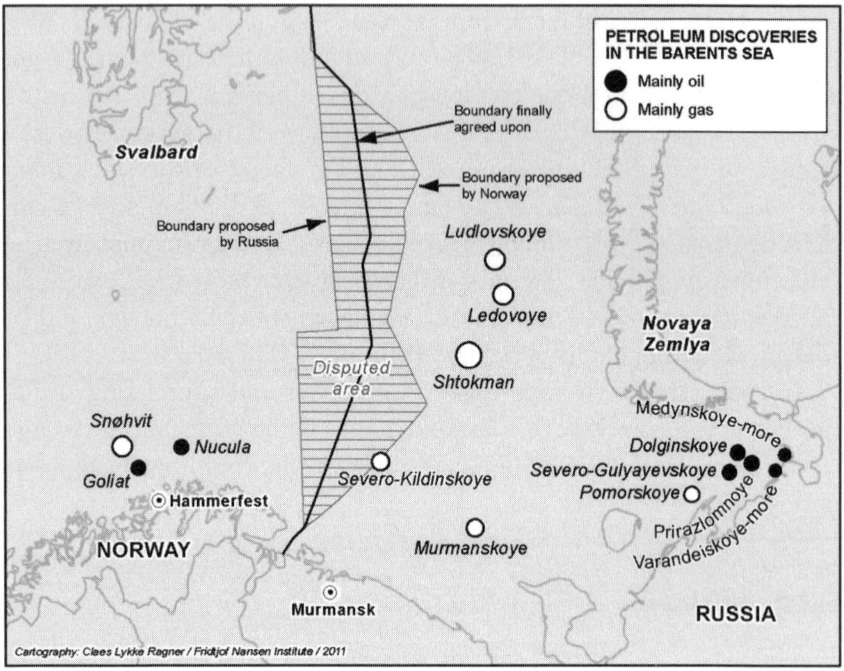

Fig. 5.2. Discoveries in the Barents Sea.
Source: Created by Claes Lykke Ragner, Fridtjof Nansen Institute.

but there are also sizeable oil resources. Although considerable uncertainty surrounds the Russian estimates, there is little doubt that the resource potential is very substantial. However, more exploration drilling is needed before development can start.

Despite these promising perspectives, the activity level in the Russian Barents Sea has remained low. A half-hearted and limited licensing round was carried out in 1999, but has attracted little interest.[10] Plans for a new comprehensive exploration program organized as a series of licensing rounds were presented in 2001 and reiterated as late as 2006, but there has been no announcement of a new round. There are several reasons for this slowness. First, the Russian petroleum industry has had many options onshore. Additionally, it has comparatively little offshore experience, and has not pressed for a more vigorous offshore program. Second, the military has been a brake, but probably less so now than earlier. Third, the Russian authorities have wanted Russian companies to be in control, but the companies have not been ready. Finally, it is also evident that the legal and tax framework has been insufficient to make risky offshore operations attractive.

Over the last five or six years, the prominence of offshore development has increased considerably in speeches made by Russian officials. This is connected to a growing realization that Russia may face a problem getting enough onshore fields on stream to maintain output, despite a huge resource base. The increased priority given to offshore fields has led to improvement in the legal framework and tax rules. However, these developments have been overshadowed by the new legislation on investment in strategic sectors that was adopted in April 2008, and accompanied changes in other laws, notably the Law on Mineral Resources. As a result, all the resources on the Russian continental shelf were declared to be of "federal significance". The continental shelf would be reserved for state companies with at least five years' work experience on work on the continental shelf — in practice Rosneft' and Gazprom.

This does not preclude joint projects with foreign companies, as long as the license remains with the Russian party. Yet, through the new legislation, the authorities ceded the initiative to the two companies. An assessment of the outlook for further petroleum development in the Russian Barents Sea must therefore take into consideration the strategies and interests of Gazprom and Rosneft more than ever before. It can be questioned whether Arctic offshore development has the same urgency for these two companies as it seems to have for the Russian state.[11] The Russian Minister of Natural Resources has expressed increased frustration over the lack of interest — on the part of these companies, and, consequently, proposals to change the law to open up for more entrants are now on the table. Several initiatives for high profile alliances for offshore exploration formed by Rosneft in 2010 and 2011 (with Chevron, ExxonMobil, and BP) can probably partly be seen as a response to the pressure from the government.

Summing up, it is evident that Russia still does not have a coherent policy for further offshore developments in the Barents Sea, or elsewhere. The prioritization of state goals is missing, while the division of functions between state organs and state companies is unclear. Private and foreign interests are kept at arm's length. A new state program for exploration and development of the continental shelf has been promised in 2011, but it remains to be seen if such a document can resolve the contradictions. There is still great uncertainty regarding the timing and pace of development, as well as the development concepts involved. The following account details major Russian fossil energy activities in the Arctic offshore.

5.2.2.1. *Shtokman*

Russia's *Shtokman* Gas and Condensate Field, located 650 km northeast of Murmansk and 540 km from shore is one of the largest offshore gas fields in the world with reserves of 3,800 bcm.[12] Again, the precise comparison with Norwegian reserve figures is difficult, since the Russian figure reflects the reserve category C1 — test-drilled (evaluated reserves), which in other cases has turned out to be overstated. However, clearly, the field is more than twice as big as the Troll Field in the North Sea, which is currently the biggest offshore gas-producing field in the world. Shtokman also contains condensate (53 million tons), which enhances its commercial attraction.

Following its discovery in 1988, seven exploration wells were drilled in the Shtokman Field during the period 1990–1996. Various international co-operation schemes for its development have been discussed over the years, but the Russian license holder, Gazprom, (through its subsidiaries) became committed to its development only after 2003 when LNG gained prominence in the company's strategy, particularly directed towards the US market. In addition, technological breakthroughs made development of the proposed world's biggest LNG project more feasible. Gazprom announced that it would develop the field in a consortium in which the largest international companies showed interest. In 2005, five companies were shortlisted for this project: Chevron, ConocoPhillips, Total, Statoil, and Norsk Hydro. Plans were changed, however, when in October 2006 Gazprom declared that foreign partial ownership was out of the question. The field would therefore be developed by Gazprom itself with the help of foreign technology providers and possibly international oil companies in assisting roles.[13]

However, this position was also soon abandoned to be followed by a new collaborative model launched in 2007. For the first phase of development, a special-purpose company, Shtokman Development AG (SDAG), was established together with Total, which now includes StatoilHydro.[14] These two foreign companies hold 25% and 24% of the shares, respectively, while the majority (51%) rests with Gazprom. The special-purpose company will develop and operate about one third of the field. It will own the infrastructure for 25 years after production start-up, after which everything will be handed over to Gazprom. Gazprom will sell the field's gas and the development license will remain with Gazprom's subsidiary, now renamed *Gazprom neft' shel'f.* The exact contractual conditions are not public and may be finally settled only in the course of 2011.

The full development of Shtokman is currently envisaged in three stages, each producing up to 23.7 bcm of gas per year, to commence at four-year intervals. Gazprom has already started planning the second phase itself to be undertaken by a newly-created and fully owned subsidiary, *Gazprom dobycha shel'f.* According to the present plans, peak production of 71.1 bcm per year will be reached after 25 years. Altogether, the field is scheduled to produce for 50 years. There are, however, expectations that it will be possible to have a fourth development phase, bringing peak production up to 95 bcm.[15]

When "New Shtokman" was launched in 2007, there was not an agreed overall technical solution for the field's development. The challenges were, and still are substantial, as the long distance to shore, drifting ice, and high waves all pose problems. Statoil and Hydro had both separately and together argued for a pure sub-sea solution referring to their experience from the Norwegian continental shelf. Gazprom, it seems, found such a solution too risky, and thus favoured a more conventional project with one or two platforms. The agreed solution is a kind of compromise: Subsea installations connected to a floating production unit (FPU). Aker Solutions, Technip, SBM Offshore and Daewoo Shipbuilding & Marine Engineering set up a consortium considered as the leading contender for a contract to construct the FPU, but in June 2011 it was announced that Aker Solutions had withdrawn from the bidding process citing unfavourable assessment of the commerciality of the project.[16]

The initial plans for the first phase indicated that half of its gas output would be transported by a new pipeline linked to the North Stream Pipeline (under construction) connecting Russia to Germany via the Baltic Sea. A smaller amount of gas would be used for local needs along the pipeline. The other half would be liquefied in the LNG plant to be constructed at Teriberka on the Kola Peninsula. The annual output of LNG would be 7.5 million tons destined for "countries in the Atlantic basin". Later phases would be dominated by LNG and total annual LNG deliveries might reach 30 million tons.[17] The official goal was originally to start deliveries of piped gas in 2013 and LNG in 2014, but the feasibility of that timeframe was questioned by many, referring to the time required for constructing the required installations and drilling.

Despite the initial announcement, the SDAG has not yet taken a firm decision to go ahead with its investments, even though it agreed, in principle, in late 2009 to do so. Whereas uncertainties relating to costs and timeframe were evident from the start of the cooperative venture, market uncertainties have taken the centre stage since 2009. The financial crisis of the last few years caused a sharp dip in the gas demand in Russia's export markets as well as within Russia itself. The need to bring in new production capacity for the sake of Russia's gas balance was suddenly less urgent than it was just a few years earlier. Nevertheless, this situation need not derail a project that is expected to come on stream well after the financial crisis is over, and demand has therefore picked up. The more serious challenge comes from the development of unconventional gas, particularly the vast resources of shale gas in the United States, which drastically changes the gas markets. North America, which has been targeted as a major market for LNG from Shtokman, may not need to import any LNG at all over the next few years; in fact, it may become a net exporter. To make the export situation even more difficult for Shtokman LNG, there may be an oversupply of LNG for some time in other potential markets, including the European market.

Thus, no wonder the SDAG announcement of 5 February 2010, declaring its postponing the investment decision to 2011, identified the uncertain market outlook as the reason.[18] The company also announced that it would decide on the related pipeline project first and then later in 2011 whether to build the LNG plant. Thus a "pipeline-only" solution is being considered. However, such a development would mean a stronger "competition" with other onshore Russian gas resources, notably those of the Yamal Peninsula, while the *raison d'ˆetre* for developing Shtokman has been its strategic location as an LNG source. Shtokman gas is commonly considered more expensive to develop than other possible sources for the integrated Russian pipeline network. Nevertheless, Shtokman could relieve some of the pressure on Gazprom's investment budget since its foreign partners will help with financing as well as management of the project. Perhaps, Shtokman as a collaborative project could also play a role in reducing some of the tensions between Europe and Russia in the gas market.

There is no consensus on how the gas market will develop in the longer term. "Gas optimists" argue that the availability of new, vast gas resources world-wide, and particularly in the United States, will change attitudes to this type of energy to make it the desired fuel for the longer term in many countries; thus, it will increase gas demand compared to the earlier estimates. This development will, in turn, support higher prices than those experienced during and in the aftermath of the recent financial crisis. Reduced attraction of nuclear power after the Fukushima incident in 2011 is also likely to increase the demand for natural gas. These are also considerations in the discussion about the start-up of a long-term project such as Shtokman.

5.2.2.2. *Prirazlomnoye*

The largest oil field identified in the Russian Barents Sea is Prirazlomnoye, located 57 km offshore from Varandey, at a depth of 20 meters. Four wells have been completed in this field since the commencement of drilling in 1989. Estimates of its exploitable reserves have, in recent years, been adjusted radically downwards to 46.4 million tons in 2010, sufficient to support an annual output of

6 million tons for about 8 years. The Rosshel'f Consortium received a license for its development in 1993. Started in 1995 as part of its development project, the construction of the Prirazlomnaya Platform at the Sevmash Shipyard in Severodvinsk was scheduled to be completed in three years. The platform is a huge steel caisson to be placed on the shallow ocean floor, like an artificial island, to contain production and storage facilities and protect the installation from the severe ice problems in the area.

However, problems soon emerged, and the construction process has come to a complete standstill several times. In addition to strictly technical problems, constant changes in the design have been a major reason for delays. It has also proved difficult to attract sufficient financial resources. Several foreign partners have been in and out of the project, including the Australian oil company BHP, the International Finance Corporation (under the World Bank) and the German company Wintershall AG. Other companies have been approached to take part in the project (e.g., notably Norsk Hydro in 2003), but they have declined, finding the project too risky and/or not commercially attractive.

In 2002, the license was transferred to Sevmorneftegaz, initially owned jointly by Gazprom and Rosneft'. Consequently, it was decided to abandon the original idea of building a complete platform in Severodvinsk. Instead, a discarded platform from the North Sea — Hutton TLP — was purchased and the topside was transported to Severodvinsk for installation on the caisson. Yet, the used platform proved to be a disaster as only very little of this purchase could be used. The topside that was finally built had to consist solely of new, imported equipment. Towards the end of 2010, the platform was towed to a shipyard in Murmansk for final works. It now seems realistic that it would be ready for production at the end of 2011 or in 2012.

In summary, the experience drawn from Prirazlomnoye has not been encouraging, but some lessons have been learnt. The development has taken longer and proven more expensive than expected.

Plans for using exclusive Russian equipment had to be abandoned. However, with the large platform in place as a hub, development of other smaller fields in the vicinity will become more attractive.

5.2.3. Norway and Russia — similarities and differences

As we have seen, there are similarities between Norway and Russia in terms of developing the Arctic energy resources, but many differences also exist. To start with, the oil and gas resource pictures in Norway and Russia differ. Whereas Russia has several unexplored areas outside the Arctic, Norway's options are more limited as it has become increasingly evident since 2000. As a result, the Norwegian petroleum industry has since strongly argued for increased activity in the North while the Russian companies have been more reluctant. In addition, both in Norway and Russia, regional interests have generally been positive towards development, but regional concerns loom larger in Norway than in Russia.

Moreover, although there are restraining forces in both countries, these also differ. In Norway, environmental considerations have slowed the general development and seriously influenced the organization of development activities. Unlike the Norwegian case, environmental restraints have so far not played any observable role in Russia's policies. This does not mean that the environment is irrelevant in the Russian case. As has been shown in other instances, environmental regulations have the potential to stop or delay industrial development, but green issues are brought in at a later stage in Russia's planning process than they are in Norway.

Unlike the case of Norway, there is a discrepancy between Russian statements and actual development of the offshore hydrocarbon sector. An important reason for this is competing policy goals. Russian companies and suppliers are supposed to play a dominant role in the development of offshore resources, but since they are not ready to play that role, their development is slow. Yet, the fact that a major project, namely Shtokman, has been under way for some time seemingly undermines the above mentioned arguments. However, in that case, the configuration of interests is different. Accordingly, the main driver for Shtokman's development has been Gazprom's interest in the LNG market. That interest has been shared by foreign, including Norwegian, companies. For this huge project, obviously regarded as strategic, the foreign companies have been willing to accept a peculiar organizational and contract structure, a structure probably inapplicable to other projects at other times.

5.3. ENVIRONMENTAL AND RESOURCE MANAGEMENT CONCERNS

The vulnerability of the Northern environment is a central issue in Norwegian thinking about the High North. Environmental concerns include the preservation of wildlife and biodiversity as well as the pristine natural surroundings, but they are also connected to specific economic interests: fisheries. The Barents Sea is one of the most bio-productive seas in the world and has very rich fishing grounds, especially for the highly valued cod. Environmental and fishery interest groups fear that pollution and spills related to petroleum activities will have a serious negative effect on biodiversity and fish resources; this may reduce catches or lower the value of fish from the Barents Sea.

For these reasons, Norway has proceeded cautiously in developing its hydrocarbon resources in the North. Concern for the environment and for the possible impact on fisheries has limited the scope of exploration in terms of spatial extent and time periods. When the Barents Sea shelf was reopened in 2003 after the halt in exploration in 2001, the "zero-discharge principle" was applied, requiring minimization of operational discharges. There is a continuing heated debate about which areas shall be opened for exploration and development. Stricter environmental regulations than elsewhere on the Norwegian continental shelf are applied to those areas where petroleum activity is allowed.

The Russian Federation shares many of the interests of Norway with regard to protection of the environment and resources in the North. It has, therefore, officially acknowledged the importance of environmental considerations. There are, however, fundamental differences between the Norwegian and Russian attitudes towards environmental challenges.

The Norwegian approach is dominated by the precautionary principle: What can be done to prevent possible problems? The debate in Norway has, to a large extent, been revolving around hypothetical situations. There have been very few actual environmental problems with the offshore activity carried out by Norway. The Russian approach is different. Its focus is on the actual situation and also on experiences from elsewhere. Accordingly, it is based on the following logic. As long as concrete problems have not been identified, like major discharges, it is difficult to turn the attention to possible future problems. "Let us wait and see what happens", is therefore a phrase commonly heard in Russia. Various explanations may exist for this attitude, but simply, many Russians, researchers and ordinary people alike, find it futile to discuss hypothetical environmental problems offshore when there are many unsolved environmental issues close to home.

The political processes surrounding environmental policy and regulations are also very different in Norway and Russia. Environmental debates are a central element in the general political discourse in Norway. In addition, several Norwegian environmental NGOs have a high profile and play an active role in such debates. Such organizations are regarded as legitimate stakeholders while even concerns expressed by laymen are considered a factor to be taken into account. On the contrary, in Russia, there is a much stronger emphasis on "expert opinions". There are, therefore, much fewer legitimate participants in the limited ongoing debate. To be fair, sometimes environmental NGOs active in Northwest Russia are consulted by the Russian authorities, if they are considered to possess certain expertise. Nevertheless, generally speaking, these NGOs are considered amateurs. Moreover, the environmental NGOs tend to focus on current, concrete issues, rather than potential future problems.[19]

Interestingly, the perceived state of environmental affairs on the Russian continental shelf has been used as an argument in favour of increased activity on the Norwegian Barents Sea shelf. Thus, in the past decade, the reasoning was geared to an assumption as to Russia's imminent large scale activities in the Barents Sea and that the Norwegian companies should be given a go ahead so that they could provide an example to the Russian industry to follow.[20] Clearly, as argued earlier in this chapter, the impression of haste on the Russian side was exaggerated, and the perception of Russian environmental regulations was superficial.

Environmental impact assessments play a central role in Norwegian petroleum policy at several stages. These are: first, before decisions about opening new areas for exploration are made, then later, before licensing rounds are announced and much later, when concrete projects are considered. Russia has a well-established system of environmental impact assessments — *OVOS* — for all kinds of industrial projects. However, impact assessments are directed towards the project level, and Russia lacks a more integrated approach to strategic planning before individual projects are developed.[21]

Against this background, there are clear conflicts between different arms of the Norwegian government regarding offshore development. The Ministry of Petroleum and Energy is generally supportive of such development, whereas the Ministry of the Environment is more sceptical. Appreciating this reality, the government tries to balance concerns and form a coherent policy. An ambitious attempt at a concerted policy was the development of a comprehensive plan for integrated management of the Barents Sea that was finalized in 2006.[22] The plan is intended to be in line with international treaties and processes stressing the need for integrated management of resources and supporting environmental considerations in sea areas. One of the most politically sensitive issues in this regard is the establishment of areas with severe restrictions — whether as Particularly Sensitive Sea Areas (PSSAs) according to IMO standards and guidelines, or other categories of marine protected areas where oil exploration will not be permitted at all. This overall management plan is an important determinant for the further opening of new areas for exploration and production.

The Russian environmental authorities are relatively weaker than their counterparts in Norway. Russia's Ministry of Natural Resources and Ecology is mainly a structure oriented towards resource exploitation. There are agencies under the ministry set up to monitor and control offshore activities. However, their political clout is not so strong. In practice, environmental policy continues to be treated as a sector interest and is given much less attention at the highest political level than Norway. Another general feature of Russian policy making relevant in this context is that issues tend not to be settled, precisely because the government does not really integrate and balance all concerns. Consequently, infighting continues even after decisions have ostensibly been made.

In some cases, Russian environmental regulations are stricter than those of Norway, e.g., with regard to discharges. Yet, questions have been raised with respect to the seriousness of *compliance* with extensive and complicated Russian laws and regulations. For example, fines imposed on perpetrators are not always high enough to deter offenders.

While Norway cannot directly affect developments on the Russian side, it can support improvements in Russian policy and regulations, a declared goal for the Norwegian government.[23] Accordingly, there is a long standing relationship and co-operation between Norwegian and Russian environmental authorities and the Russian side has shown interest in adopting elements from the Norwegian integrated management plan in the Russian Barents Sea. As a positive example, increased co-operation between the Norwegian and Russian research institutions led to publication of a joint environmental status report for the Barents Sea in 2009.[24] The next phase might be the establishment of common monitoring systems. However, the weakness of Russian environmental authorities has placed limitations on this effort. The Norwegian dream of developing common principles for resource and environmental management for the whole Barents Sea is still far from realization, even if the general Russian attitude to co-operation on resources management is positive.[25]

In the Norwegian domestic debate, climate concerns are being used as an argument against extension of petroleum activities in the North. The reasoning is, that as rapid climate change in the Arctic is becoming more and more evident, increasing petroleum activity in that region symbolizes wrong priorities in national energy policy. The government does not agree with this and holds that climate change is a global problem and that oil production in the Arctic is not different from such production in other regions. Measures to combat climate change should, therefore, be applied where they are most effective. Thus, the policy is to gradually develop Arctic resources, but at the same time employ various measures to reduce domestic emissions as well as work toward strong international agreements. On the other hand, climate change is not used as an argument in support of more Arctic petroleum activity. The almost euphoric statements about the access to new acreage provided by the receding ice-cover heard in some international media is almost absent in the Norwegian public. This may be explained by a degree of political sensitivity in the Norwegian oil industry, but also by the fact that there is no ice cover in the parts of the Norwegian Barents Sea opened for petroleum activities so far.

On the Russian side, climate change and climate politics have much lower importance on the government agenda compared to that of fossil energy exploration. Increased open waters due to Arctic melting are sometimes referred to, but they are not major issues in official documents about development strategies in the North.

5.4. JURISDICTIONAL ISSUES

According to the UN Convention on the Law of the Sea (UNCLOS), the coastal states enjoy sovereign rights to exploit the minerals in the seabed. However, the continental shelf is not sovereign territory. Other states enjoy full rights to shipping activities and as a rule would be granted permission for scientific activity. Yet, clearly, the coastal states may have large economic benefits from their continental shelves, provided, of course, they contain valuable and exploitable resources.

Article 76 of the UNCLOS grants all coastal states a continental shelf of 200 nautical miles if it does not meet another state's shelf.[26] Nevertheless, the continental shelf may also extend further out as stipulated by the Convention. Accordingly, "The continental shelf of a coastal State comprises the seabed and subsoil of the submarine areas that extend beyond its territorial sea throughout the natural prolongation of its land territory to the outer edge of the continental margin".[27] The delimitation of continental shelves to bordering shelves, as well as its outer delimitation is a complicated issue as discussed below.

5.4.1. *Outer delimitation of continental shelf*

The convention describes in some detail geological requirements for including a seabed area in the "extended" continental shelf, entailing serious scientific documentation. A special commission (The Commission on the Limits of the Continental Shelf) comprising technical specialists has been set up by the UN to review submissions of such documentation. If the commission finds the documentation sufficient, it will endorse, which means international recognition of the outer limit.

With respect to the Arctic, Norway submitted its documentation in 2006 after several years of preparatory work. Accordingly, Norway claimed an extended shelf — beyond 200 nautical miles — in three separate sea areas in the Arctic Ocean, the Barents Sea and the Norwegian and Greenland Seas, totalling approximately 235,000 square km (see Fig. 5.3). The claims were accompanied with extensive geological documentation.[28] The commission reviewed the documentation, requiring some small adjustments.[29] Other states did not object to the submission as such. However, certain states (Russia, Denmark, and Iceland) formally noted that final delimitation would be pending the resolution of boundary delimitation disputes with neighbouring shelves. Norway clearly agreed with this formula as also Spain, which noted that acceptance of the Norwegian submission did not prejudice its position regarding maritime zones around the Svalbard Archipelago.

The commission delivered its final recommendation on the Norwegian submission in March 2009. This was the first submission to be finally processed and meant that Norway now had internationally recognized delimitation of its extended continental shelf.

In fact, there is so far only one other state that has submitted documentation of an extended continental shelf in the Arctic to the Commission, namely Russia. Russia submitted its documentation on 20 December 2001, including the Fig. 5.4. In effect, Russia claimed sovereign rights over resources on the seabed area of some 1.2 million km^2 outside the 200-nautical-mile line. Its geological argument was based on the linkage between the claimed areas' ridges on the seabed to Russia's mainland.[30] All the other four Arctic coastal states as

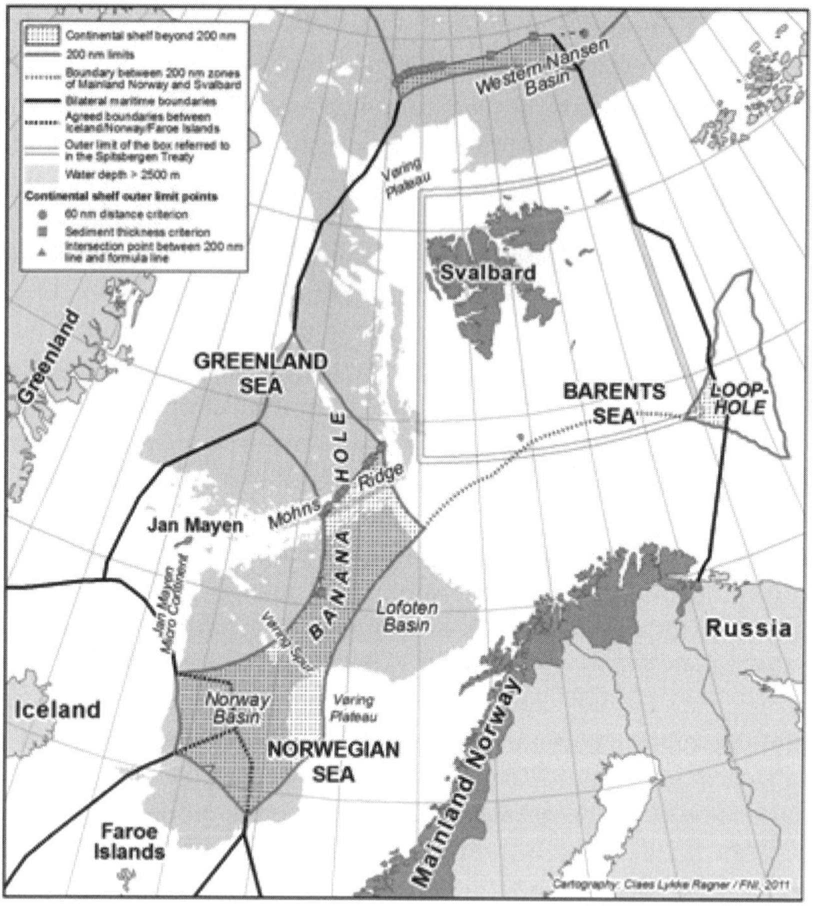

Fig. 5.3. Norwegian Continental Shelf Boundaries.
Source: Created by Claes Lykke Ragner, Fridtjof Nansen Institute.

well as Japan questioned elements in the Russian submission.[31] The Commission found the substantiation of the Arctic claim insufficient and asked for more information.[32] Since then a new submission has been under preparation, reportedly to be finalized by 2013. Comprehensive research expeditions are now being organized by the Russian authorities to collect data for the Russian claim. Interestingly, there have been several instances of data exchange and scientific cooperation with other Arctic states in this effort.[33]

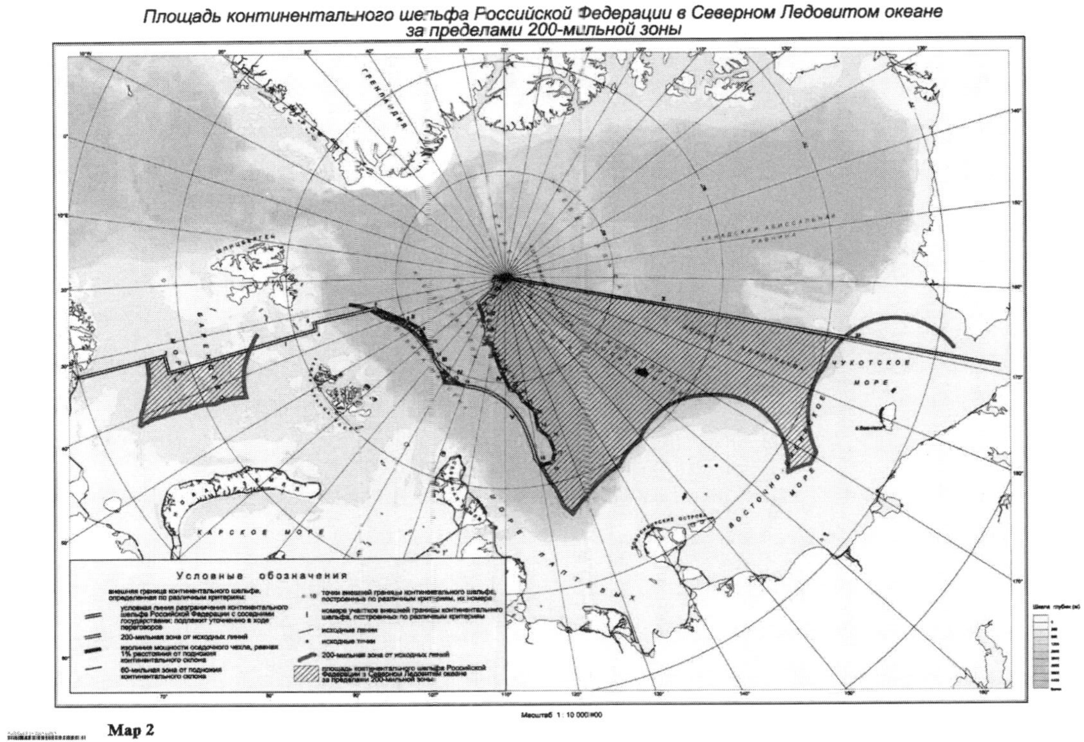

Fig. 5.4. Continental Shelf Area of the Russian Federation in the Arctic Ocean beyond the 200-Nautical-Mile Zone.

Source: Russia's submission to the Commission on the Limits of the Continental Shelf 2001. http://www.un.org/Depts/los/clcs_new/submissions_files/submission_rus.htm.

Despite these efforts, what will happen if Russia's documentation is again found insufficient? The answer to this question is naturally of great significance for international relations in the Arctic and raises the possibility of international conflict. We will return to this issue after discussing the recent important developments in Arctic jurisdictional issues — the delimitation of the Barents Sea.

5.4.2. *Delimitation of neighbouring continental shelves: the case of the Barents sea*

For more than 40 years, Norway and Russia/the Soviet Union disagreed over their maritime boundary in the Barents Sea and the Arctic Ocean. Norway first asked for negotiations concerning the delimitation of the continental shelf in 1967, and in 1970, the two countries discussed the issue for the first time in informal meetings. Formal negotiations started in 1974. Norway and Russia asserted overlapping claims to an area in the Barents Sea and the Arctic Ocean of some 175,000 square kilometres. Norway based its claim on the "median line" principle, delimiting the border equidistant from the nearest points of the two countries' coastline. This median line is the Eastern boundary of the disputed area in Fig. 5.2.

The Soviet Union, like Norway, also referred to the Continental Shelf Convention of 1958, but claimed "special circumstances" — widely defined, including population and military considerations — for deviation from the median line. It also pointed to a 1926 decree, which claimed that all islands within a triangle from the outer borders to the North Pole belonged to the USSR. It was argued that the decree was relevant to the EEZ and the continental shelf too. Negotiations were characterized by deadlock in long periods, and there were no strong signs that a resolution was imminent.

The settlement of the dispute in 2010 came as a surprise to Norway and Russia. It divided the Area of Overlapping Claims, or the disputed area, into two nearly equal parts.[34] In essence, Russia has now come to terms with a compromise that had been contemplated early on by the Norwegian side. Accordingly, explanations for the outcome of the negotiations must be sought primarily on the Russian side.

Both Norwegian and Russian leaders have emphasized that resolving this delimitation of the continental shelf is the culmination of a long, gradual process and that the line now drawn is based on international law. The emphasis on the legality of the delimitation according to the international law was reflected in the remarks of the two parties' foreign ministers commenting on the settlement. According to the Norwegian Foreign Minister, Jonas Gahr Støre: "The line has been computed in relation to the relevant coasts on either side, on the basis of modern principles of international law",[35] and in the words of Russia's Foreign Minister, Sergey Lavrov: "both the principles enshrined [in the agreement] and the parameters of the delimitation described by the agreement to a full extent correspond with the norms of international law".[36] It is interesting to know that the original Russian position partly based on the sector decree, declared initially as "nonnegotiable", was weakened with Russia's accession to the UNCLOS in 1997. According to the UNCLOS, delimitation of adjacent continental shelves and exclusive economic zones "shall be effected by agreement on the basis of international law, as referred to in Article 38 of the Statute of the International Court of Justice, in order to achieve an equitable solution, thus not pointing to any specific principle."[37] The International Court of Justice, which in a series of decisions has crystallized what constitutes relevant circumstances, has narrowed their application to permanent, natural characteristics of neighbouring territories, in particular, the length of the respective coastlines.[38] Thus, on this basis, and with reference to court decisions, a series of points were defined, ultimately making up a borderline further west than a "pure" application of the median line principle would have given, nevertheless, much further east than the original Russian claim. The legal development and the dynamics of the negotiations are important to understand *how* the line was drawn, but a willingness to find a political compromise must be added as an explanation as to *why* a solution was arrived at.

Russia and Norway's bilateral relations have steadily improved on the background of constructive co-operation — especially in the management of fish stocks. The prospects of Norwegian–Russian petroleum co-operation is another factor with positive overtones that helps us understand why it was *politically* feasible to reach a compromise in 2010. A generally good atmosphere must be seen as beneficial and probably a precondition for the agreement in its current form and timing. Nevertheless, the deal cannot be understood only in a Norwegian–Russian context as broader developments and policy goals were also contributing factors.

The previously disputed area (PDA) is believed to hold large oil and gas resources. This potential is often used to explain why it has been so difficult to reach an agreement. There are big uncertainties about the resource potential though, since no drilling has been undertaken. However, both Norway and Russia have carried out some seismic surveying.[39] Russian officials have, on several occasions, stated that they expect the area to be resource rich and have even produced estimates. After the delimitation deal was signed, Russian Natural Resources Minister Yuriy Trutnev stressed the uncertainty of the estimates, but also announced that there were up to ten structures, which *might* have resources in the category very large (more than 1 trillion cubic meter [tcm] of natural gas or 1 billion tons of oil — also known as giants) or *unique* (more than 5 tcm of natural gas or 5 billion tons of oil — also known as super giants).[40] Many Norwegian commentators have taken it for granted that the potential for petroleum development was a decisive factor in reaching an agreement. Surely, in the official comments by President Dmitry Medvedev, for instance, energy loomed large. However, these statements, and also the official comments from the Russian Ministry of Foreign Affairs, were of a fairly general nature acknowledging that the area might contain significant resources and expressing a wish to cooperate with Norway in this regard. Nevertheless, they did not reveal a concrete Russian intention to explore and develop the area. It does not seem that energy development was a strong driver for reaching a solution, and lobbying from Russian energy companies to open and develop the area was not evident. That is not surprising given that until now there has not been any heavy-weight Russian industrial actor with a strong interest in venturing into the PDA. As mentioned earlier, legislation adopted in 2008 reserved new projects on the Russian continental shelf for state companies — in practice Rosneft' and Gazprom, which, until now, have not prioritized expensive Arctic offshore exploration.[41]

After the agreement was adopted and much of the secrecy lifted, there has been no sign of expansive plans for the area, particularly, a hitherto secret plan for its rapid development. This does not mean that an assessment of the area's hydrocarbon potential was not part of the decision, as reflected in official comments. However, these comments treated energy development more as an opportunity that could be exploited once the resolution was found. Therefore, concrete *energy-industrial* considerations were probably not a driver for the deal. Yet, if there was no intention to develop resources now, the natural question would be: What else could be gained from signing the deal now?

I argue that there is good reason to connect the deal with broader Russian interests in the Arctic. Russia has, in fact, explicitly made this link. It was a major explanation or justification for the agreement made by Russia's Ministry of Foreign Affairs and also in unofficial remarks. According to the ministry's spokesman, "The delimitation is essential also from the point of view of progress for our claim in the Commission on the limits of the continental shelf".[42] He was referring to the recommendations from the Commission to Russia in 2002 of the need to arrive at an agreement with Norway to establish the outer limits of the Russian continental shelf.[43] However, Norway received final recommendations regarding its claim in 2009, so the pending delimitation line did not block Norway's submission from being dealt with. Thus, it may be questioned what impact an unresolved border would have had on the Russian claim. When the Russian officials are stressing this connection, it could be because only the slightest risk of disturbing the process in the commission would be a weighty argument for reaching an agreement in the Barents Sea.

The delimitation agreement in the Barents Sea also has a more indirect link with Russia's claim in the continental shelf commission, however, in both cases the question is whether disputes should be solved by way of unilateral actions and conflict or through international law and negotiations.

In the Barents Sea deal, both Russia and Norway were very clear about the delimitation line being influenced by developments in international law — as practiced by the International Court of Justice. The mere fact that the parties found an orderly and peaceful solution to one of the most prominent jurisdictional disputes in the Arctic, with potentially vast resources at stake, demonstrates that conflict is not inescapable. Nevertheless, perhaps, more important than this, it seems that Russia has concluded that it is best served by peaceful negotiations embedded in international law, rather than using pressure and linkages against a small neighbour (as sometimes feared in Norway). Such reasoning does not imply that Russia has taken on an altruistic stance in international relations, but that it sees its interests best served through exploitation of international law and institutions, as spelled out in some of its foreign policy documents. Russia's willingness to find a compromise solution with Norway can be interpreted as an expression of Russian interest in keeping the Arctic peaceful and avoiding conflicts. It thus shows that there are sufficient instruments available to solve disputes and there is no need to introduce new institutions. In other words, supporting and reinforcing the UNCLOS constitutes the basis for securing Russia's resource rights in the Arctic. Russian Prime Minister Vladimir Putin referred to the delimitation agreement as an example showing that "existing Arctic problems, including the continental shelf questions, can be solved in the spirit of cooperation, through negotiations on the basis of existing international legal norms".[44]

Russia has much to lose and nothing to gain by weakening the UNCLOS's authority. Hence, any statements or actions that can weaken it should be avoided. Resolving the delimitation issue in apparent compliance with international law would be the right thing to do in this perspective. Thus, general support to the UNCLOS is an important factor explaining Russia's willingness to compromise in the Barents Sea.[45] In fact, President Medvedev made this point explicit when he submitted the treaty with Norway to the Duma in February 2011: "Maritime delimitation between Russia and Norway strengthens legal regulations in the Arctic based on the UNCLOS of 1982, in addition to creating an atmosphere of peace and cooperation in the region."[46]

5.5. OUTLOOK

In the case of the outer delimitation of the continental shelf in the Arctic, Russia faces some of the same choices as it does in the Barents Sea, but with even greater ramifications — unilateral moves and the risk of conflict or a peaceful, orderly process. However, the question remains what will happen if a new Russian submission is not accepted. Several outcomes of the process in the Commission on the Limits of the Continental Shelf are conceivable. One extreme would be for Russia to just declare unilaterally that its continental shelf reaches up to the North Pole. The country could simply withdraw from the UNCLOS. It would still retain the right to a continental shelf, and would find itself in the same position as the United States, which stands outside the convention, and thus would have to rely on customary law to support its claim. However, this approach would provide a much less secure legal position than a recommendation from the commission would, which offers final and binding limits.

The unilateral approach obviously bears the risk of conflict with the other Arctic states. Will Russia be willing to take such a risk? How much is the absence of conflict, and the preservation of a cooperative regime in the Arctic, worth to Russia? It is, in my opinion, likely that strong nationalistic voices would support such unilateralism. Yet, Russia's official policy has, since the ratification of UNCLOS, undoubtedly adhered to the UNCLOS framework. Moreover, as concluded above, Russia has a lot to lose if it undermines the authority of international law and the UNCLOS, in particular, in the Arctic. Consequently, it will seek to avoid a conflict situation since a conflict, even if not armed, would prove to the world that the institutions set up under the UNCLOS do not work, and this would further weaken the legitimacy of the Convention. In the current situation with a changing balance of power globally, as alluded to above, such weakening would be very dangerous for Russia.

The other extreme outcome if a second submission is returned would be for Russia to accept that the continental shelf does not reach as far as claimed and come back with a revised, less expansive position. This alternative would definitely show respect for international law. Nonetheless, all the same, such a stance would involve large domestic political costs. Any Russian leader would find it very difficult to abandon Russia's ambitious Arctic claim, and especially to explain such abandonment to the Russians.

I will argue that a third outcome is also possible, and indeed the most likely. That option is postponement. First, there is the possibility of several rounds with the commission. That could take decades. However, even if, at the end of the day, it is clear that claims in the Arctic cannot be reconciled or substantiated, all the Arctic states may see it in their interest to leave it like that, i.e., agree on disagreements and go on with their business. Apart from a shared interest in preserving the UNCLOS in the Arctic, a realistic assessment of economic interests would convince them that a conflict is not worthwhile. The seabed areas, which may be contested, are after all very deep down. It seems unlikely that industrial activity there can become profitable in the foreseeable future, requiring perhaps many decades for that outcome. The most authoritative and much cited assessment of Arctic mineral resources, from the US Geological Survey, maintains that most resources are likely to be found in relatively shallow waters, within the 200-mile limit.[47] Most of these uncontroversial continental shelves are virtually unexplored. That should logically happen first.

5.6. CONCLUSION

Even if Norway and Russia have carried out exploration drilling in the Barents Sea since the early 1980s, only one field — Snøhvit in the Norwegian part of the sea — has so far been put into production. Because of decreasing production in the North Sea, there is now a growing Norwegian interest in the Arctic shelf. Recent discoveries have fuelled optimism. However, environmental concerns have placed some limitations on development. On the Russian side, exploration has been limited, but many discoveries have been made. The Russian effort has been characterized by irreconcilable goals and insufficient framework conditions, thus progress has been very slow. A major gas development on the Russian side — Shtokman — has been underway for some time involving both Norway's Statoil and French Total together with Russia's Gazprom. A final investment decision is promised in 2011. A go-ahead for that project will be of great significance for further Russian–Norwegian co-operation. Nevertheless, changes in the conditions for operations on the Russian continental shelf allowing for broader access to private Russian as well as foreign companies are required to speed up development. Based on the background of recent political debates in Russia, such changes seem likely.

The delimitation agreement between Russia and Norway in the Barents Sea opens a new large section of the continental shelf for petroleum activities, although energy is not seen as a major driver for reaching the agreement in 2010. For Russia, the deal should be seen in relation to Russia's broader Arctic policy and interests, where stability and support for the UNCLOS are central goals.

The UNCLOS provides coastal states with the right to a continental shelf beyond 200 nautical miles if the connection to the mainland can be substantiated. The risk that Russia's extensive claim may be in conflict with the claims of other Arctic states or that it will not be generally recognized has made several observers fear open conflict. This is by no means a given, however, taking into account the Arctic states' shared interests in avoiding conflict as well as the limited economic value for the foreseeable future of the seabed areas that might be contested.

The energy resources in the Arctic likely to be explored and developed are located on the continental shelves firmly under the jurisdiction of the Arctic coastal states. However, the development and transportation of these resources will involve companies from several countries and areas under various jurisdictions. Practical solutions that consider the environmental challenges will require intensified international cooperation.

BIBLIOGRAPHY

Arneberg, P, O Korneev, O Titov, JE Stiansen (eds.), A Filin, JR Hansen, Å Høines, and S Marasaev (co-eds.) (2009). Joint Norwegian-Russian environmental status 2008 Report on the Barents Sea Ecosystem. IMR/PINRO Joint Report Series.

Baker, B (2010). Law, science and the continental shelf: The Russian federation and the promise of Arctic cooperation. *American University International Law Review*, 25(2), 252–281.

Barents Observer (2011). 21 June. http://www.barentsobserver.com/aker-solutions-backs-out-of-shtokman-contract.4935600.html (Another contending group is composed of Saipem, Samsung and Sofec).

Christian, JL (2007). Petroleum discourse in the European Arctic: The Norwegian case. *Polar Record*, (43), 247–254.

Facts — the Norwegian Petroleum Sector 2009 (2009). Ministry of Petroleum and Energy and Norwegian Petroleum Directorate, pp. 138–139.

Gahr Støre, J (2010). Most is North — The High North and the way ahead — an international perspective. Lecture at the University of Tromsø, 29 April. http://www.regjeringen.no/en/dep/ud/Whats-new/Speeches-and-articles/speeches foreign/2010/Most-is-north.html?id=602113 Official translation

Gautier, D *et al.* (2009). Assessment of undiscovered oil and gas in the Arctic. *Science*, (324).

Gazprom (2010). Совет директоров «Штокман Девелопмент АГ» утвердил дальнейшие направления развития Штокмановского проекта. ("The Board of Directors: 'Shtokman Development AG' Approved Future Direction of the Shtokman Project.") Press release 5 February. http://www.gazprom.ru/press/news/2010/february/article 76114/

Geir, H, JH Jørgensen and A Moe (2007). 'Miljøpersepsjoner i Nordvest-Russland: Problemoppfatninger knyttet til petroleumsutbygging i Barentshavet. ("Environmental Perceptions in North-Western Russia: Perspectives on petroleum development in the Barents Sea.") *Internastional Politikk*, (65), 7–22.

Integrated Management Plan Ready (2006). Office of the Prime Minister, 31 March. http://www.regjeringen.no/upload/MD/Vedlegg/Svalbard%20og%20polaromraadene/Forvaltningsplan%20Barentshavet/PDF008-0506 engelsk-TS.pdf.

Interview with Yuri Komarov, General director of Shtokman Development AG (2008). "Shtokman Ostaetsya Rossiyskim Proektom." *Gazprom*, 3.

Jensen, Ø (2010). Towards setting the outer limits of the continental shelf in the Arctic: On the Norwegian submission and recommendations of the commission. In *Law, Technology and Science for Oceans in Globalisation*, Vidas, Davor (ed.), pp. 521–538. Leiden/Boston: Martinus Nijhogff Publishers/Brill.

Jensen, Ø (2011). Current legal developments — the Barents Sea: Treaty between Norway and the Russian federation concerning maritime delimitation and cooperation in the Barents Sea and the Arctic ocean. *Journal of Marine and Coastal Law*, 26(1), 151–168.

Lavrov, SV (2011). Press conference, Russian Ministry of Foreign Affairs, 13 January. http://www.mid.ru. (Author's translation)

McDorman, TL (2010). The outer continental shelf in the Arctic ocean: Legal framework and recent developments. In *Law, Technology and Science for Oceans in Globalisation — IUU Fishing, Oil Pollution, Bioprospecting, Outer Continental Shelf*, Vidas, Davor (ed.), pp. 497–520. Leiden/Boston: Martinus Nijhoff Publishers.

Mineral'nye resursy rossiyskogo shel'fa (2006). *Mineral'nye resursy Rossii — Ekonomika i uprevalenie*, Special (ed.), p. 16.

Ministry of Environment. Report 1 to the Storting 2009–2010 *The State budget — Ministry of the environment*, Chapter 10.

Moe, A and AK Jørgensen (2000). Offshore mineral development in the Russian Barents sea. *Post Soviet Geography and Economics*, (41), 98–133.

Moe, A and EW Rowe (2009). Northern offshore oil and gas resources: Policy challenges and approaches. In *Russia and the North*, Elana Wilson Rowe (ed.), pp. 107–128. Ottawa: University of Ottawa Press.

Moe, A (2006). "Sjtokman-beslutningen: Forklaringer og Implikasjoner." ("The Shtokman Decision: Explanations and Implications"). *Nordisk Østforum*, (20), 389–403.

Moe, A (2010). Russian and Norwegian Petroleum Strategies in the Barents Sea. *Arctic Review on Law and Politics*, 1(2), 225–248.

Moe, A (2011). Russia's Arctic continental shelf claim: A slow burning fuse? In *Geopolitical and legal aspects of Canada's and Europe's Northern Dimensions*, M Nuttall and AD Nuttall (eds.). Virginia: CCI Press.

Moe, A and EW Rowe (2009). Northern offshore oil and gas resources: Policy challenges and approaches. In *Russia and the North*, E Wilson Rowe (ed.), pp. 107–128. Ottawa: University of Ottawa Press.

Moe, A, D Fjærtoft and Øverland (2011). Space and timing: Why was the Barents sea delimitation dispute resolved in 2010? *Polar Geography*, 34(3), 145–162.

Nyland, B (2011). New Spring in the Barents Sea. *Norwegian Continental Shelf*, 1, p. 3. http:www.npd.no/en/publications/norwegian-continentalshelf/no1-2011/new-spring-in-the-barents-sea–/

Putin, VV (2010). Speech at the International Forum The Arctic — territory of dialogue. Moscow, 23 September. http://premier.gov.ru/events/news/12304/ (accessed 10 February 2011).

RIA Novosti (2010). 15 September.

Russian MFA Spokesman Andrei Nesterenko Response to Media Question Regarding Russian-Norwegian Agreement on Maritime Delimitation in the Barents Sea and the Arctic Ocean, 28 April 2010. Official translation. http://www.mid.ru (accessed 20 January 2011)

Russian-Norwegian Treaty on Maritime Delimitation has been submitted to the State Duma for ratification (2011). 10 February. http://eng.kremlin.ru/news/1758 (official translation).

Solodyankina, S and J Koeppel (2009). The environmental impact assessment process for oil and gas extraction projects in the Russian federation: Possibilities for improvement. *Impact Assessment and Project Appraisal*, (27), 77–83.

Statement by the Chairman of the Commission on the Limits of the Continental Shelf on the progress of work in the Commission (2009). CLCS/62, 20 April. http://www.un.org/Depts/los/clcs new/commission recommendations.htm

Statement by the Chairman of the Commission on the Limits of the Continental Shelf on the progress of work in the Commission (2002). Commission on the Limits of the Continental Shelf, Tenth session, New York, CLCS/32, 25 March–12 April. http://daccess-dds-ny.un.org/doc/UNDOC/GEN/N02/327/68/PDF/N0232768.pdf?OpenElement

Statoil on course with Snohvit LNG (2010). *Upstreamonline.com*, 19 April.

Tamnes, R (1997). *Oljealder 1965–1995 — Norsk utenrikspolitisk historie bind 6*, pp. 323–324. Oslo: Universitetsforlaget.

UN. *United Nations Convention on the Law of the Sea of 10 December 1982.* http://www.un.org/Depts/los/convention agreements/convention overview convention.htm

United Nations General Assembly (2002). Fifty-seventh session, Agenda item 25 (a) Oceans and the Law of the Sea — Report of the Secretary-General. Addendum A/57/57/Add. 18 October. http://daccess-dds-ny.un.org/doc/UNDOC/GEN/N02/629/28/PDF/N0262928.pdf?Open Element

United Nations General Assembly, Fifty-seventh session, Agenda item 25 (a) Oceans and the law of the sea — Report of the Secretary-General (2002). "In the case of the Barents and Bering Seas, the Commission recommended to the Russian Federation, upon entry into force of the maritime boundary delimitation agreements with Norway in the Barents Sea, and with the United States of America in the Bering Sea, to transmit to the Commission the charts and coordinates of the delimitation lines as they would represent the outer limits of the continental shelf of the Russian Federation extending beyond 200 nautical miles in the Barents Sea and the Bering Sea respectively." Addendum A/57/57/Add.1, 8 October. http://daccess-dds-ny.un.org/doc/UNDOC/GEN/N02/629/28/PDF/N0262928.pdf?OpenElement [accessed 13 January 2011].

ENDNOTES

1 Research for this chapter has been carried out within the framework of the "Geopolitics in the High North" project and the RussCasp project, both financed by the Research Council of Norway. The chapter draws on Moe, A (2010). Russian and Norwegian Petroleum Strategies in the Barents Sea. *Arctic Review on Law and Politics*, 1(2) and uses elements from Moe, A (2011). Russia's Arctic continental shelf claim: A slow burning fuse?' In *Geopolitical and legal aspects of Canada's and Europe's Northern Dimensions*, M Nuttall and AD Nuttall (eds.). Virginia: CCI Press.

2 Russia is a significantly larger producer than Norway, being the world's No.1 gas producer and alternating between being the world's No.1 or 2 oil producers. In terms of exports they are more similar: Russia is the world's No. 1 gas exporter and Norway No. 2. Russia is the world's second largest oil exporter, whereas Norway has been No. 6 in recent years.

3 Tamnes, R (1997). *Oljealder 1965 1995 — Norsk utenrikspolitisk historie bind 6*, pp. 323 324. Oslo: Universitetsforlaget.

4 Nyland, B (2011). New Spring in the Barents Sea. *Norwegian Continental Shelf*, 1, pp. 3. http://www.npd.no/en/publications/norwegian-continental-shelf/no1-2011/new-spring-in-the-barents-sea–/

5 Statoil on Course with Snohvit LNG (2010). *Upstreamonline.com*, 19 April.

6 Ministry of Petroleum and Energy and Norwegian Petroleum Directorate (2009). *Facts — the Norwegian Petroleum Sector 2009*. pp. 138–139.

7 Mineral'nye resursy rossiyskogo shel'fa (2006). *Mineral'nye resursy Rossii–Ekonomika i uprevalenie*, Special (ed.), p. 16.

8 *Ibid.* p.18.

9 Note that these reserve/resource classifications are not directly comparable. Norwegian "remaining reserves" constitute a more strictly defined category than Russian "recoverable resources".

10 Moe, A and AK Jørgensen (2000). Offshore Mineral Development in the Russian Barents Sea. *Post Soviet Geography and Economics*, (41), 98–133.

11 Moe, A and EW Rowe (2009). Northern offshore oil and gas resources: Policy challenges and approaches. In *Russia and the North*, Elana Wilson Rowe (ed.), pp. 107–128. Ottawa: University of Ottawa Press, Ottawa.

12 http://www.gazprom.ru. Bcm — billion cubic meters — are used here when referring to Russian data. Note that a Russian cubic meter is measured differently than a standard cubic meter, the measure used on e.g., the Norwegian continental shelf. Russian figures must be multiplied by 0.91 to get the equivalent in standard cubic meters. The number here only includes the license plot. The field extends somewhat beyond the plot.

13 Moe, A (2006). Sjtokman-beslutningen: Forklaringer og Implikasjoner (The Shtokman Decision: Explanations and Implications). *Nordisk Østforum*, (20), 389–403.

14 The merged company later retook the name Statoil.

15 Interview with Yuri Komarov, General director of Shtokman Development AG (2008). "Shtokman Ostaetsya Rossiyskim Proektom," *Gazprom*, 3.

16 *Barents Observer* (2011). 21 June. http://www.barentsobserver.com/aker-solutions-backs-out-of-shtokman-contract.4935600.html (Another contending group is composed of Saipem, Samsung and Sofec).

17 Interview with Yuri Komarov, General director of Shtokman Development AG (2008). "Shtokman Ostaetsya Rossiyskim Proektom," *Gazprom, 3, op. cit.*

18 Gazprom (2010). Совет директоров «Штокман Девелопмент АГ» утвердил дальнейшие направления развития Штокмановского проекта ("The Board of Directors: 'Shtokman Development AG' Approved Future Direction of the Shtokman Project"). Press release 5 February. http://www.gazprom.ru/press/news/2010/february/article76114/

19 Geir, H, JH Jørgensen and A Moe (2007). "Miljøpersepsjoner i Nordvest-Russland: Problemoppfatninger knyttet til petroleumsutbygging i Barentshavet," ("Environmental Perceptions in North-Western Russia: Perspectives on Petroleum Development in the Barents Sea"). *Internastional Politikk*, (65), 7–22.

20 Jensen, LC (2007). Petroleum discourse in the European Arctic: The Norwegian case. *Polar Record*, (43), 247–254.

21 Solodyankina, S and J Koeppel (2009). The environmental impact assessment process for oil and gas extraction projects in the Russian federation: Possibilities for improvement. *Impact Assessment and Project Appraisal*, (27), 77–83.

22 Office of the Prime Minister (2006). Integrated Management Plan Ready, 31 March. http://www.regjeringen.no/upload/MD/Vedlegg/Svalbard%

20og%20polaromraadene/Forvaltningsplan%20Barentshavet/PDF0080-506engelsk-TS.pdf.

23 Ministry of Environment. Report 1 to the Storting 2009–2010 *The State Budget — Ministry of the Environment*, Chapter 10.

24 Arneberg, P, O Korneev, O Titov and JE Stiansen (eds.), A Filin, JR Hansen, Å Høines and S Marasaev, (co-eds.). Joint Norwegian-Russian environmental status 2008 Report on the Barents Sea Ecosystem. IMR/PINRO Joint Report Series.

25 In a press release from the Russian Ministry of natural resources 11 December 2009 after a meeting with the state secretary in the Norwegian ministry of petroleum and energy, a series of co-operation areas for the near future was listed, including exploration and production of hydrocarbons on the Arctic and Barents continental shelf.

26 UN, *United Nations Convention on the Law of the Sea of 10 December 1982*. http://www.un.org/Depts/los/convention agreements/convention overview convention.htm

27 *Op. cit.* Article 76.

28 Jensen, Ø (2010). Towards setting the outer limits of the continental shelf in the Arctic: On the Norwegian submission and recommendations of the commission. In *Law, Technology and Science for Oceans in Globalisation*, Davor Vidas (ed.), pp. 521–538. Leiden/Boston: Martinus Nijho? Publishers/Brill.

29 Statement by the Chairman of the Commission on the Limits of the Continental Shelf on the progress of work in the Commission (2009). CLCS/62, 20 April. http://www.un.org/Depts/los/clcs new/ commission recommendations.htm

30 McDorman, TL (2010). The outer continental shelf in the Arctic ocean: Legal framework and recent developments. In *Law, Technology and Science for Oceans in Globalisation — IUU Fishing, Oil Pollution, Bioprospecting, Outer Continental Shelf*, Davor Vidas (ed.), pp. 497–520. Leiden/Boston: Martinus Nijhoff Publishers.

31 Statement by the Chairman of the Commission on the Limits of the Continental Shelf on the progress of work in the Commission, Commission on the Limits of the Continental Shelf (2002). Tenth session, New York, CLCS/32, 25 March–12 April. http://daccess-dds-ny.un.org/ doc/UNDOC/GEN/N02/327/68/PDF/N0232768.pdf?Open Element

32 United Nations General Assembly (2002). Fifty-seventh session, Agenda item 25 (a) Oceans and the Law of the Sea — Report of the Secretary-General. Addendum A/57/57/Add. 18 October. http://daccess-ddsny.un.org/doc/UNDOC/GEN/N02/629/28/PDF/N0262928.pdf?Open Element

[33] Baker, B (2010). Law, science and the continental shelf: The Russian federation and the promise of Arctic cooperation. *American University International Law Review*, 25(2), 252–281.

[34] Jensen, Ø (2011). Current legal developments — the Barents Sea: Treaty between Norway and the Russian Federation Concerning Maritime Delimitation and Cooperation in the Barents Sea and the Arctic Ocean. *Journal of Marine and Coastal Law*, 26(1), 151–168.

[35] Gahr Støre, J (2010). Most is North — The High North and the Way Ahead — an International Perspective. Lecture at the University of Tromsø, 29 April. http://www.regjeringen.no/en/dep/ud/Whats-new/Speeches-and-articles/speeches
foreign/2010/Most-is-north.html?id= 602113 Official translation

[36] Lavrov, SV (2011). Press conference, Russian Ministry of Foreign Affairs, 13 January. http://www.mid.ru. (Author's translation)

[37] UN, *United Nations Convention on the Law of the Sea of 10 December 1982.* Article 83. http://www.un.org/Depts/los/convention agreements/ convention overview convention.htm

[38] Jensen, Ø (2011). Current Legal Developments — the Barents Sea: Treaty between Norway and the Russian Federation Concerning Maritime Delimitation and Cooperation in the Barents Sea and the Arctic Ocean. *op. cit.*

[39] Norway ceased seismic surveys in 1976, but USSR continued well into the 1980s. Thereafter a moratorium on all forms of exploration was mutually respected.

[40] RIA Novosti (2010). 15 September.

[41] Moe, A and E Wilson Rowe (2009). Northern offshore oil and gas resources: Policy challenges and approaches. In *Russia and the North*, E Wilson Rowe (ed.), 107–128. Ottawa: University of Ottawa Press.

[42] Russian MFA Spokesman Andrei Nesterenko Response to Media Question Regarding Russian-Norwegian Agreement on Maritime Delimitation in the Barents Sea and the Arctic Ocean, 28 April 2010. Official translation. http://www.mid.ru [accessed 20 January 2011]

[43] United Nations General Assembly, Fifty-seventh session, Agenda item 25 (a) Oceans and the law of the sea — Report of the Secretary-General (2002). "In the case of the Barents and Bering Seas, the Commission recommended to the Russian Federation, upon entry into force of the maritime boundary delimitation agreements with Norway in the Barents Sea, and with the United States of America in the Bering Sea, to transmit to the Commission the charts and coordinates of the delimitation lines as they would represent the outer limits of the continental shelf of the Russian Federation extending beyond 200 nautical miles in the Barents Sea and the Bering Sea respectively," Addendum A/57/57/Add.1,

8 October. http://daccess-dds-ny.un.org/doc/UNDOC/GEN/N02/629/28/PDF/N0262928.pdf?OpenElement [accessed 13 January 2011].

44 Putin, VV (2010). Speech at the International Forum The Arctic — territory of dialogue (Moscow) 23 September, http://premier.gov.ru/events/news/12304/ [Accessed February 10, 2011].

45 A more extensive treatment of factors explaining the agreement, expanding on the discussion in this chapter, can be found in Moe, A, D Fjærtoft and I Øverland (2011). Space and Timing: Why was the Barents Sea Delimitation Dispute Resolved in 2010? *Polar Geography*, 34(3), 145–162.

46 Russian-Norwegian Treaty on Maritime Delimitation has been submitted to the State Duma for ratification. 10 February 2011. http://eng.kremlin.ru/news/1758 (official translation).

47 Gautier, D *et al.* (2009). Assessment of undiscovered oil and gas in the Arctic. *Science*, (324).

SECTION III

WHAT ARE THE INTERESTS FOR THE NON-REGIONAL LARGE ECONOMIES?

6

THE ARCTIC: GEOPOLITICS, INTERNATIONAL RELATIONS AND ENERGY SECURITY — A VIEW FROM INDIA

Neil Gadihoke

6.1. INTRODUCTION

We are living in an era of unprecedented scientific research and technological development. Across domains and subjects, mankind has taken giant strides, powered by a quest for discovery. Human beings have been to the moon, we have sent missions to Jupiter and Pluto, discovered new galaxies and have searched incessantly for primary evidence of life on other planets. However, in a paradox almost too poetic to comprehend, humans seem to have made little progress in deciphering a curious phenomenon closer home — that of the melting of the desolate, frigid and mysterious Arctic on our very own planet Earth.

Two decades ago, a young scholar Oran Young penned a prescient article "The Age of the Arctic",[1] in which he proclaimed: "Today, the Arctic is rapidly becoming a focus for defence and development issues that touch on the core interests of each of the superpowers." Young then believed that the world was "entering the age of the Arctic, an era in which those concerned with international peace and security will urgently need to know much more about the region and in which policy makers in the Arctic rim states will become increasingly concerned." At that stage, not many were ready to believe his prophesy that soon the Arctic Ocean would be the modern-day equivalent of the Mediterranean of ancient times.

Various scientific studies project that Arctic ice will be dramatically reduced or possibly disappear during part of the summer as early as 2050. Conservative estimates suggest that a 12%–40% reduction in summer ice has already occurred.[2] Shrinkage of the polar ice cap would open fabled sea routes to Asia that European explorers had sought in vain for centuries. Commercially viable Arctic sea lanes, which have the potential to make the Arctic Ocean a major global trade route, are anticipated to be opened for part of the year well before 2050.[3] Therefore, as envisioned by Oran Young, we are on the cusp of the "Age of the Arctic". Also, this region is said to hold potentially as much as 25% of the remaining undiscovered global reserves of oil and gas, with the largest reserves expected to be found in the Russian part. Other types of fossil energy, minerals and metals are also abundant such as coal, nickel, copper, tungsten, lead, zinc, gold, silver, diamonds, manganese, chromium, and titanium. A dispassionate analysis of such trends suggests that this phenomenon will have a substantial impact on Asian geopolitics, and, by implication, India.

Consequently, the strategic thinking community in India, as across the world, is analysing and speculating on the economic and geopolitical consequences of the melting of the Arctic Ocean. The warmer climate, new technologies for the extraction of hydrocarbons and the increasing interest in Northern maritime routes are igniting a new great game in the Arctic. These dynamics, in the next few decades, will introduce competition for the secure access to energy, minerals, food and markets on a scale hitherto unknown to the region. Though the Arctic region seems far away from India, the progress of the "Polar Game" will certainly have a bearing on India's futuristic strategic and economic environment. This is owing to the fact that India is a fairly large country in geographic terms, is the second most populous nation and is the second fastest growing economy after China. In sum, it is too large a stakeholder in the international system to escape the consequences of a rapidly melting Arctic. It may, therefore, be time for India to start contemplating an

Arctic strategy, which encompasses issues far beyond India's ongoing benign activities like scientific and oceanographic research.

6.2. THE ARCTIC MELT: EXTENT, INTENSITY AND IMPLICATIONS

In order to grasp the potential impact on India, it would be prudent to briefly delve into the present status of the Arctic melt, as well as the likely future scenario. This would then obviously lead to the macro level implications, both positive and negative, for the entire international community.

On average, over the last 20 to 30 years, the Arctic ice cover has shrunk by around 7% per year in summer, and 2% per year in winter; and in recent years, the pace has accelerated.[4] The area of open water in the Arctic is increasing rapidly in summer, and the ice cover is getting younger, thinner and softer by the year. The question, then, is: in the future, how fast will the melting occur? The extent of global warming, in the coming decades, is very hard to predict. However, if the melting will proceed as fast as it has on average over the last 50 years, the summer sea ice could be gone before the end of this century. Based on a continuation of the more rapid melting that has occurred in the last 5–10 years alone, with additional periods of abrupt melting, recent studies conclude that the Arctic Ocean could have virtually ice-free summers by 2040.[5] Periods of radical abrupt melting could set in as early as 2015, due to threshold effects related to increased inflow of warm water. However, even with the summer ice cap gone, the winter ice cap would still be extensive.

6.2.1. *The impact of an ice-free Arctic*

At a macro level, there are three important positive implications of an Arctic meltdown. First, the navigation season along the Northern Sea Route (NSR) and the North-West Passage (NWP) will gradually increase from the current 20–30 days per year, and, at some point, it will be possible to sail almost straight across the North Pole. Second, as the ice retreats, access to many of the raw material resources in the Arctic area will become easier. Further, the higher temperatures will mean increased productivity of fish stocks and forests in the Arctic, likely leading to increased exploitation.

The possibility of these three positive implications is governed by four negative caveats. Warming in the Arctic will include greater ice movement and wave action, which will increase the risks to sailing and maritime operations there. Second, rising sea levels and river flows in the Arctic will contribute significantly to a rise in sea levels across the globe. This is the most significant implication for India. Third, an Arctic meltdown could cause significant changes in global weather patterns, which would, in turn, exacerbate the consequences of global warming. Fourth, the fragile ecosystem of the Arctic region would be damaged due to increased human exploitation and pollution.

6.2.2. *Implications for India*

For India, the three most important repercussions of the opening-up of the Arctic are potentially shorter inter-continental shipping routes, the likely improved access to vast energy and mineral resources and the consequences of rising sea levels.

6.2.2.1. *The Arctic highway*

This Polar Sea is a centrally placed "Mediterranean" ocean that connects the North Atlantic and the Pacific Oceans and constitutes, by far, the shortest distance between the two. For centuries, therefore, explorers had searched for a viable NSR between the Atlantic and the Pacific Oceans without much success. The sea routes that were finally discovered, namely the North East Passage, or the NSR as it is most often called, off the coast of Siberia, and the NWP between the Canadian Arctic Islands, were only partly open, at best a few short weeks per year; they never came to play any role in transcontinental commercial shipping. Due to the heavy costs and hostile environment, the direct route "over the top" was hardly even contemplated, even with the advent of the modern ice breakers, and nuclear powered submarines, with their ability to cross the ice covered Polar Sea safely and with relative ease. If they become a reality, the Arctic shipping routes, will potentially mean a 40% reduction in the distance from northern Europe to northeast Asia and the northwest coast of North America, compared to southerly sea routes via Suez or Panama.[6] As the Arctic ice-free season gets longer, and the ice during the covered periods becomes easier to penetrate, a northern sea lane linking Europe and Asia will become increasingly attractive to shippers, as it may substantially reduce the time and costs of transportation between the Atlantic and the Pacific Oceans.

In sum, the Arctic shipping routes could cut the cost of a single voyage by a large container ship by as much as 20%, saving the shipping industry billions of dollars a year.[7] The savings would be even greater for the megaships that are unable to fit through the Panama and Suez Canals and thus currently sail around the Cape of Good Hope and Cape Horn. Moreover, these Arctic routes would also allow commercial and military vessels to avoid sailing through politically unstable Middle Eastern waters and the pirate-infested Horn of Africa and South China Sea.

India is according the highest priority to development of its maritime sector, premised on its location, overlooking the dense sea lanes of the Indian Ocean. To this end, the new Maritime Agenda 2010 2020 (MA 10–20), launched by Indian Prime Minister Manmohan Singh on 13 January 2011 in New Delhi, with a budgetary allocation of INR 500,000 crores, is a seminal document. The MA 10–20, in the form of a study and plan, compares the Indian maritime sector to global standards in order to highlight the deficiencies in India. It outlines the trends in international shipping and port infrastructure and makes future projections so as to identify the specific areas of improvement required in India's maritime sector. Thus, the MA 10–20 provides a roadmap for the development of the maritime sector and sets the targets for the relevant departments to achieve. The key targets include creating a port capacity of 3,200 million tons for the handling of some 2,500 million tons of cargo, improving port performance to be on par with the best in the world, increasing tonnage under the Indian flag and increasing India's share in global shipbuilding to 5%.

Given such endeavours, the Arctic melt, with its potential to divert the shipping traffic away from the Indian peninsula, will need to get factored into India's long term maritime development plans. This is owing to the fact that the container volumes and shipping loads, handled by India's present and future ports, which are astride the main east-west sea transportation lanes in the Indian Ocean Region, may decrease for four months a year.

6.2.2.2. *Arctic region as a petroleum province*

The US Geological Survey and the Norwegian company StatoilHydro estimate that the Arctic holds as much as 25% of the world's remaining undiscovered oil and gas deposits.[8] Some analysts estimate the Arctic hydrocarbon reserves at 40% of the world's total oil and gas reserves.[9] The Arctic Ocean's long, outstretched continental shelf is another indication of the potential for commercially accessible offshore oil and gas resources. As the global oil prices continue to remain volatile, the commercial exploitation of the Arctic hydrocarbon resources has become technologically and economically feasible. The advantage of Arctic resource exploitation is the fact that it is situated in a region where conflict and political instability do not threaten secure and reliable delivery. Therefore, the Arctic oil and gas could radically alter energy import patterns, particularly for countries like China, Japan and South Korea that are increasingly dependent on oil and gas from distant and politically volatile regions like the Middle East and Africa. Further, the Arctic's energy potential could also impact India. To this end, this chapter briefly elucidates the future energy dynamics of India and China.

Currently the world's 11th largest economy (fourth in terms of purchasing power parity), India could occupy the third slot after the United States and China in 25–30 years, if she manages to sustain her economic growth rate. Since such growth is directly related to energy consumption, India's thriving manufacturing, transport, and agricultural sectors are expected to drive energy demand at an average annual rate of 3–4% over the next quarter century. India, presently fifth on the energy consumption ladder, is expected to overtake Japan and Russia to reach the third spot by 2030. Investments worth billions of dollars have been made in overseas hydrocarbon assets extending from Sakhalin in the Russian Far East across Central Asia and Africa to South America.[10] India has evinced interest in the Arctic's fossil reserves and the largest Indian oil company — the Oil and Natural Gas Videsh Limited (OVL) — has approached Rosneft, a major state-run Russian energy company, for exploration in the Yamal Peninsula in the Russian portion of the Arctic.

China's dependence on imported energy is more acute. By the mid-1990s, China had moved from being a minor and largely self-sufficient energy consumer, to become the world's fastest growing energy importer and a significant player in global energy markets. Until then, China's energy demand was met by domestic sources. In 1993, China's oil production and consumption were approximately 3 mb/d (million barrels per day). However, while its production increased slightly, reaching just 3.7 mb/d in 2008, its consumption doubled during the same period. Consequently, the Chinese government has identified uninterrupted supplies of energy as being a major factor in its comprehensive national power capability. The Chinese perceive that the narrow Malacca Straits, through which a substantial portion of the Chinese oil imports from the Middle East and Africa pass, could be interdicted by adversarial naval forces. This concern was echoed in 2003, when President Hu Jintao declared "certain major powers were bent on controlling the Strait, and called for the adoption of new strategies to mitigate the perceived vulnerability."[11] Since then Chinese officials and media have begun to portray the "Malacca dilemma" (discussed further below) as a pertinent security concern for China. According to the *China Youth Daily*, "it is no exaggeration to say that whoever controls the Straits of Malacca will also have a stranglehold on this energy route.[12]

In sum, China, which is vitally dependent on the Straits of Malacca for most of its energy supplies, would be inclined to promote this emerging alternative route. China is already looking north and engaging with Iceland, which is strategically located in the Atlantic–Arctic route to China. This route assumes more importance in the light of China's fast accelerating dependence on oil from countries on the west coast of Africa — most notably Angola, Nigeria, Gabon, and Equatorial Guinea. The vital question arising out of China's use of the Arctic route is the implications that such a development would have for India. It is well known that India and China fought a border war in 1962 and thus the lingering mistrust of Chinese military expansion is pervasive in security circles in India.

Experts acknowledge that given the relative geographical location of air bases and army formations on both sides of the India–China border, a military face-off — if it were ever to occur — would be evenly matched and likely result in a stalemate. The Indian Ocean may, therefore, become the arena wherein maritime forces would play a decisive role. China's huge energy imports travelling on sea lanes in the Indian Ocean, whether originating from the Middle East or Africa, make her vulnerable. In this context, some believe that India needs to be prepared at sea, and must leverage her geographical advantage and exert pressure on China's energy transport vulnerabilities. However, if the Chinese energy transport vulnerabilities are reduced due to the opening of the Arctic, China could become more assertive not only with India, but with other countries in the region, and this would impact regional security and attendant geopolitics.

Though the potential of the Arctic route to resolve its Malacca dilemma is high, it is pertinent to mention that the Malacca dilemma could instead become a "Bering Straits" dilemma subject to more focused strategic leverage by the Arctic rim states, all of which India enjoys excellent relationships.

6.2.2.3. *Sea level rise*

Over the last 140,000 years, the global sea level has varied over a range of more than 120 meters.[13] It stabilized over the last few thousand years and there was little change between about 1 AD and 1800 AD. In the 19th century it began to rise again and it accelerated again in the early 20th century.[14] Satellite altimeter measurements show a rate of rise of about 3 mm per year since the early 1990s. Thermal expansion is producing about half the current 3 mm per year increase in global sea levels. The contribution of this factor has increased from around 0.5 mm per year over the second half of the 20th century to around 1.6 mm per year over the last 12–14 years.[15]

The impact of the Arctic melt on the global sea level rise can neither be specifically co-related nor quantified. In the Arctic Circle, the main concern is the land-based ice like the Greenland Ice Sheet, which would raise sea levels by a few meters if it completely melted. Insofar as the ice in the Arctic sea is concerned, there is a consensus that since it is already floating on the ocean, it will not raise sea levels when it melts, but this does not also imply that its impact on the global sea level is nil. For one thing, less sea ice means that the ocean will absorb sunlight that the ice would have reflected away. Thus the sun will warm the water, which is most dense at 4°C (its density decreasing both above and below this temperature) and as the overall water temperature increases, the water will expand, making the oceans rise.

The retreating Arctic sea ice plate is turning into open water, which absorbs far more solar radiation, and thus heats up more. This dynamic is creating a self-reinforcing melting cycle. Consequently, the rise in the global sea level rise, which is presently occurring at a noticeable rate, could well spiral out of control were any unforeseen "wild card" phenomena to happen. Further, the Arctic melt contributes substantially to global warming in another way as well. The permafrost (permanently frozen soil) on the land frontiers in the polar region is shrinking, thus releasing large amounts of trapped greenhouse gases (mainly methane), which have aggravated global warming. In a nutshell, the Arctic's melting does contribute to rising sea levels, the impacts of which are numerous, unanticipated and deleterious, especially for a developing third world country like India.

Another major contributing factor to the rising sea levels is the melting of glaciers, the polar ice cap, and the Greenland and Antarctic ice sheets. This is believed to produce about one third or more of the current 3 mm per year annual increase in the global sea level. The complete melting of the Greenland ice sheet alone could increase global mean sea levels by around 7 meters.[16] It would probably take about 1,000 years, but it is believed that melting ice from Greenland could still contribute significantly to rising sea levels over the next 50–100 years.[17] Further, global warming is causing the atmosphere to contain more water vapour, which enhances greenhouse effects and leads to more warming and melting in a self-reinforcing cycle.

Insofar as the impact of rising sea levels on the Indian coast is concerned, a short case study of a coastal village in the Indian state of Orissa is presented in the following paragraphs.

6.2.2.4. *Satabhaya village, Orissa*[18] *— a short case study*

The continuing rise of the Bay of Bengal has progressively inundated the coastal village of Satabhaya in Orissa's Kendrapa District. This hamlet village has been forced to move inland thrice over the last 20 years. Two decades ago, the sea was a scenic 2-km walk from the village. Today, the original village location is about 1 km offshore. This hamlet is symbolic of the dangers of rising sea levels, which could lead to a multitude of detrimental factors, the chief among which are:

- ***Forced Migration*** Of the two hundred families that once lived in Satabhaya and its neighbouring Kanhapur Village, only 32 remain. In response, the state government has gradually shrunk the revenue map of the area.
- ***Increased Salinization*** A few hundred meters away from the Satabhaya village are lush paddy fields and forests, but villagers fear they may not remain lush for long. Saline water is seeping under their fields and into the sand covering, shrivelling their crops. Further, the rising sea will increase salinity in the surrounding coastal areas, dramatically reducing food crop yields and jeopardizing drinking water supplies.

6.2.3. *Vulnerabilities in the Indian maritime neighbourhood*

Most nations that are most vulnerable to rising sea levels do not have the resources to prepare for it. In South Asia, India, Bangladesh, Pakistan and Sri Lanka have large populations living in "at-risk" coastal areas. Small island nations such as the Maldives are at severe risk because they do not have enough land at higher elevations to support displaced coastal populations.[19] Another challenge for coastal populations is the danger of losing their fresh-water supplies as rising sea levels push saltwater into their aquifers. For these reasons, those living on several small island nations like the Maldives could be forced to evacuate over the period of the 21st century. Specific vulnerabilities in India's maritime neighbourhoods are elucidated in the succeeding paragraphs.

Bangladesh is the world's third most vulnerable country with regard to rising sea levels in terms of the number of people and is in the top ten in terms of percentage of population living in low-lying coastal zones.[20] Currently, almost 40 million people live in the coastal areas of Bangladesh. Loss of coastal land to the sea in this vulnerable zone — now predicted to reach up to 3% by the 2030s, 6% in the 2050s, and 13% by 2080 — is likely to generate a steady flow of displaced people to India.[21] Owing to the sealing of the land border with India, a large portion of the displaced population may resort to the sea route for illegal migration.

The Maldives holds the record for being the lowest country in the world with a maximum natural ground level of only 2.3 meters above sea level and an average height of only 1.5 m above sea level. In November 2008, Maldivian President Nasheed announced plans to look into purchasing land in India, Sri Lanka, and Australia due to the possibility of much of the islands being inundated by rising sea levels.[22] The purchase of land is intended to be made from a fund generated by tourism. Some reports suggest that Islamic fundamentalism is finding its way into this Islamic nation, albeit on a small scale. One likely scenario is that the forced relocation of the population, loss of economic opportunities in tandem with religious fundamentalism some years down the line, could create a national security crisis in the Maldives, which in turn would have an adverse impact on India's maritime security.

India's Lakshwadeep group of islands has problems similar to that of the Maldives, insofar as terrain and vulnerability are concerned. Whatever be the consequences of climate change in the Maldives, it will be replicated in these islands.

The British Indian Ocean Territory — the island of Diego Garcia — is a critical staging facility for US and British naval and air forces operating in the Middle East and Central Asia. It sits just a few feet above sea level at its highest point. According to a 2007 Washington think-tank report (by an eminent panel of 11 retired US Admirals and Generals), the Diego Garcia base — because of its low average elevation of 1.3 meters above sea level — is considered the prime example of a "losing place" in the face of global sea level rise over the next three to four decades.[23] When coupled with the recent unrest in the Middle East, especially in Bahrain, which is a major US Navy base in the Indian Ocean Region, the search for alternate military bases by the US has the potential to influence regional geopolitics in a substantial fashion.

6.2.4. *Complication of maritime boundary disputes*

Maritime boundaries are determined by a baseline or low water mark. Any shift in baseline points can alter boundaries significantly and become a source of contention. Almost any change in sea levels could have remarkable effects on certain baselines; these, in turn, could have potentially dramatic consequences on maritime boundaries. Rising sea levels in particular geographic situations are likely to result in significant shifts in the outer boundaries of the oceanic zones claimed by coastal states. Even submergence of small atolls, rocks, and low-lying islands could have implications for maritime claims.

This phenomenon has the potential to aggravate the already existing complications in maritime issues between neighbouring states.

As a case in point, India has outstanding maritime boundary disputes with Pakistan and Bangladesh. The zone of contention with Pakistan is the delimitation of maritime boundaries in the Sir Creek area. As for Bangladesh, the outstanding issue is of the delimitation of the maritime boundary which has implications for a substantial portion of an exclusive economic zone (EEZ), believed to be rich in gas. Any shift in baseline points, due to rising sea levels, would alter the boundaries significantly and could become a source of greater contention.

Further, the Indian strategic thinking community will need to devote deep thought to issues such as: If the Maldives are submerged, will it lose its seat in the UN? Will its current EEZ revert to international waters? How vulnerable are India's island territories of Andaman and Nicobar Islands and Lakshawadeep Islands to sea level rise? Consequently, will India be a net gainer or net loser in terms of EEZ, due to rising ocean levels? These and other such matters related to submergence of low-lying land could be crucial foreign policy issues of the future.

6.2.5. *Operational tasking of India's maritime forces*

Higher mean sea levels along the Indian peninsula would increasingly result in coastal inundation and soil salinity. More significantly, they would compound the surges from cyclones. Therefore, the growing risk of natural disasters and crisis situations arising from climate change, especially in India's maritime zone, will increase the pressure on naval forces to undertake regular disaster relief and humanitarian aid operations. Consider the following: the Indian Navy was extensively deployed for relief operations following the havoc unleashed by the unprecedented flooding of Mumbai city in July 2005, which was caused by a combination of incessant rain and high tides (precluding rain water from being drained out). Were the high tide to become "permanent" (due to rising sea levels), what would be the impact on Mumbai during the monsoons? Apart from such crises, the maritime security forces will also have to contend with the increasing volume of illegal 'climate' migrants from Bangladesh/Maldives resorting to the sea route. Further, the Indian maritime forces will have to carry out vulnerability assessments of its bases and coastal installations against degradation due to extreme weather and rising sea levels. The construction of new jetties, dry docks and infrastructure close to the sea will require monitoring of the rate of sea level rise.

6.2.6. *India's Arctic linkages and Arctic jurisdiction issues*

India's linkages to the Arctic date back to the early part of the last century when it signed the 'Treaty between Norway, the United States of America, Denmark, France, Italy, Japan, the Netherlands, Great Britain and Ireland and the British overseas Dominions and

Sweden concerning Svalbard'on 9 February 1920 in Paris and which entered into force in August 1925.[24] The Treaty is also referred to as the "Treaty concerning the Archipelago of Spitsbergen" or the "Svalbard Treaty". At that time, India was part of the British overseas dominions.

In the present era, India's first permanent station in the Arctic region, named *Himadri*, was inaugurated on 1 July 2008. It is located in the Norwegian territory of Ny-Alesund on Spitsbergan (Svalbard). The National Center for Antarctic and Ocean Research (NCAOR) of India works under the Ministry of Earth Sciences and controls and manages the Indian Arctic Center and its related activities. The NCAOR has already signed a memorandum of understanding on polar research with the Norwegian Polar Institute.

It must be understood that, in the past, territorial expansion into the Arctic has taken take place without the kind of conflict and strife that has characterized expansion of national sovereignty in the world's temperate zones. Even when there were clashes of national interests,[25] states normally did not permit them to develop into major conflicts. The Arctic was not worth the risk. The main reason was that the states for a long time regarded the Arctic as having limited economic value.

However, in the present epoch, the economic potential of the Arctic resources and their viability for exploitation as a result of the melting ice, has ensured that the days when Arctic states, easily and almost absent-mindedly, would draw boundaries, sell (Alaska) or give away (Svalbard) territories,[26] *are irrevocably over*. Five different countries border the Arctic, but only two international borders, namely the Russian–Norwegian and the US–Canadian, run overland, and none of them is disputed. There is, however, disagreement concerning the extension of coastal state sovereignty in the Polar Sea. There is also disagreement over Norway's claim to the continental shelf and the EEZ around Svalbard, the delineation of the continental shelf between Norway and Russia in the Barents Sea and between Canada and the US in the Beaufort Sea. In addition, the maritime border between Denmark and Canada in the Nares Straits between Ellesmere Island and Greenland is disputed.

Such issues, if allowed to fester, will have a wide impact on various regional and non-regional countries, including India. The situation is especially dangerous because there are currently no broad political agreements or comprehensive legal structures that can provide for the orderly development of the region or mediate political disagreements over Arctic resources or sea lanes. There are few legal frameworks that offer guidance. The Arctic Council formed in 1996 exists to address environmental issues,[27] but has remained silent on security concerns.

Many observers argue that the United Nations Convention on the Law of the Seas (UNCLOS) is the correct tool to manage the thawing Arctic.[28] The convention provides mechanisms for states to settle boundary disputes and submit claims for additional resources beyond their EEZ. Furthermore, though the UNCLOS sets aside the resources in the high seas as the common heritage of humankind, it allows states bordering ice-covered waters to enforce more stringent environmental regulations. It also defines which seaways are the sovereign possessions of states and which international passages are open to unfettered navigation. However, UNCLOS cannot be seamlessly applied to the Arctic. The region's unique geographic circumstances do not allow for a neat application of this legal framework. To this end, there is a need to upscale both the charter and the membership of the Arctic council so that it is empowered to tackle all emerging issues, many of which have impacts on distant geographies, like India's maritime neighbourhood.

6.2.7. *The Arctic council membership — India's case*

The US Secretary of State, Hillary Clinton, during the March 2010 Arctic summit said:

> *I hope the Arctic will always showcase our ability to work together, not create new divisions. What happens in the Arctic will have broad consequences for the earth and its climate. The melting of sea ice, glaciers and permafrost will affect people and ecosystems around the world, and understanding how these changes fit together is a task that demands international co-operation.*[29]

Non-Arctic states, like India, have a sturdy case for participating as observers in the Arctic Council. The reasons are numerous. On a macro level, the Arctic is a common heritage of mankind and belongs to all the people on the planet. Further, non-Arctic states' have established research stations, dispatched research vessels and also instituted polar research studies in their respective countries on climate change, ecology, environment, human habitat, sustainable resource exploitation, geological studies and new shipping routes. India has a sophisticated scientific research station 'Himadri' at Ny Alesund and its scientists have undertaken seven expeditions to the Arctic.

The decisions made by the Arctic powers will profoundly shape the future of not only the regional situation, but also the future of many distant stakeholders, who, because of their size, will not be able to escape the consequences of the Arctic melt. India, owing to its size, population, growing economy, and democratic credentials can be expected to play a constructive role in the debate and discussions related to the Arctic. Additionally, for rapidly developing countries such as China and India, which are at the energy intensive phase of their growth, the desire to exploit hydrocarbons, marine living resources and seabed minerals is high on their agenda. However, a substantial proportion of these resources lies under the national jurisdiction of the Arctic states and can be exploited only through international collaboration. Further, as the Arctic melts, it can be expected to be a busy place with fishing fleets from far and wide visiting the area. Therefore, it is important to make the Arctic Council more inclusive and multilateral so that the perspectives of both Arctic and non-Arctic stakeholders can be taken into cognizance. Most importantly, the Arctic melt impacts the lives of people living in the tropics in an adverse manner due to rising sea levels. Therefore, these people must have a voice in the governance of the Arctic. In sum, there exists a strong case for India to be accorded observer status in the Arctic Council, as an initial step.

6.2.8. *Climate change and the respective emerging geopolitics*

Across the globe, a series of studies are presently underway to deepen the understanding of the relationship between climate change and national security. In September 2009, the Central Intelligence Agency

(CIA) established a "Center on Climate Change and National Security" as the focal point for its work on the subject. This institution is led by senior specialists from the Directorate of Intelligence and the Directorate of Science and Technology.[30] Its charter is not to look solely at the science of climate change per se, but at the national security consequences of phenomena such as rising sea levels, population shifts, and heightened competition for natural resources.[31]

It is with this background of global research and "crystal ball gazing" that the geopolitical dynamics in the Arctic need to be viewed. As a case in point, a Russian expedition promoted and blessed personally by Vladimir Putin, planted the tricolour flag of Russia four kilometres under the North Pole in 2007. It caused much consternation in the international community and was seen by some strategic analysts as the opening move in the great polar game. A primary objective of the Russian expedition was seen to be collection of data to support Russia's territorial claims over the Lomonosove and Mendeleyev Ridges.

Just days after the Russian explorers planted their flag on the Arctic seabed, the US government launched its own expedition. The political fallout over the Russian expedition prompted the Canadian government to launch a "sovereignty operation", known as Operation Nanook, in the Canadian Arctic. Nanook consisted of two surface ships, a submarine and 700 military personnel performing manoeuvres in Nunavut, Frobisher Bay, the Hudson Straits and the Davis Straits. The Canadian government's swift reaction to developments in the Arctic reflected the perceived increasing importance of Arctic issues.

For India, a few uncertainties emerge from the fluidity of the Arctic situation which consequently pose a few questions. As a case in point, there is the view that the competing territorial claims and immense potential of untapped energy sources, will mean that increased militarization of the Arctic may be inevitable. The Arctic sovereignty issues may result in concerned countries diverting their naval forces to the Arctic, with a consequent diminished presence in other parts of the globe. For example, if some of the US Navy is diverted from the Indian Ocean Region (IOR), which regional Navies are likely to fill the power vacuum? This is an issue that will mandate deep strategic thought in India. On another level, as the Arctic becomes a geopolitical pivot between great powers, the hydrocarbon resources there will have significant strategic implications for India and China. China is already the world's second largest consumer of hydrocarbon resources and India is catching up fast. If there are increased military tensions arising from governments' desires to gain jurisdictional control over Arctic resources, the nations in the region will surely adopt aggressive postures. The disagreements, if not resolved, could spiral into disputes 'analogous to the Spratly Islands' disputes where numerous states claim sovereignty, but no lucid picture of ownership exists.

In considering such probabilities, cognizance will have to be taken of the present realities. The counterpoint to increased militarization of the Arctic is the fact that, compared to merchant vessels, passage through the Arctic by warships is hazardous. The hulls and propulsion systems of the latter, as well as their intricate superstructures, are more susceptible to icing, thus jeopardizing their stability and sea-keeping qualities. There also exists the possibility of the Arctic saga unfolding into a complimentary approach by the Arctic rim states for the Arctic resources, which will permit a suitable environment for trans-Arctic trade and also for resource exploitation. Only time will tell how the scenarios play out.

The sustainable, large scale availability of the Arctic hydrocarbon resources to Europe, North America and East Asia would certainly reduce the geopolitical importance and significance of the Middle East. Whether, the reduced global focus and loss of oil revenues in the decades ahead would lead to political instability in the Middle East is a major issue which India must consider in its strategic calculus. This is owing to the fact that even with the availability of oil in the Arctic, India will continue to be dependent on oil from the Middle East.

Nicholas Spykman once famously said: "Geography is the most fundamental factor in foreign policy because it is the most permanent."[32] In the geopolitical discourses in the Western world, the Arctic has been seen to play a critical role, only as the Northern wall that enclosed the erstwhile Soviet Union and now its successor state of Russia. There would certainly be consequences if this frontier were to gradually diminish and additionally endow the already energy-rich Russia with even larger energy reserves. Whether this will lead Russia to become the new "Saudi Arabia" is a question still up in the air. If the Arctic "melt" does revive the sagging military potential of Russia, it may conceivably have an impact on the rising Chinese power.

In sum, there are a plethora of Arctic imponderables — "uncertainties" inherent in any "future" — all of which give rise to many geopolitical questions. At this stage, though they may be left unanswered, the peril will be greater if they are left unasked.

6.3. CONCLUSION

Climate change is not a theoretical hypothesis, but an irrefutable fact. The projections about its future implications vary, but sufficient scientific data exists to conclude that the magnitude and, more importantly, the speed of climate change pose a frightening challenge to mankind.

Insofar as the Arctic region is concerned, there are indications that the ice cover is declining much faster than previously estimated. Studies predict that the entire Arctic Ocean could be ice free in summer (at least for a short period of time), much sooner than what present calculations suggest. The continuing and rapid decrease in multi-year sea ice in the central Arctic Ocean will improve the possibility of marine navigation in all seasons. A combination of technical and economic feasibility will be decisive in deciding the timing of the opening of the trans-Arctic sea route. The Arctic meltdown's most direct bearing on India is on its maritime domain, specifically, its contribution to the global sea level rise, the impacts of which can be serious and unforeseen for a developing third world country like India, which has a long coastline and high population densities in its littorals.

Clearly, the consequences of the Arctic meltdown are transnational and transoceanic in nature, thus, there is a need to involve major players, even if geographically distant, in the deliberations of the Arctic's future. It is imperative to evolve a framework of deliberations that will ensure an orderly and collaborative approach in extracting the untapped energy underneath the ice cap and in managing the shipping through the area. There is a strong case for widening the scope of the Arctic Council's charter and increasing its members. India has the resources and influence to contribute positively to the evolving Arctic situation. To this end, it may broaden co-operation with the Arctic nations, establish bilateral dialogues and discussions to understand the developing politico-strategic positions in the Arctic region, and participate in Arctic resource assessment and exploitation studies. In addition, regular expeditions to the Arctic to consolidate scientific research and develop technological capabilities to exploit the Arctic's living and nonliving resources need to be engendered.

In conclusion, though the melting of the Arctic, at first glance, appears to be a phenomenon that is too distant to have an impact on India, a deeper analysis of available evidence suggests it has the potential for a substantial impact on Asian geopolitics, and, by implication, impinge on India's security and energy calculus. Therefore, the need of the hour is for the Indian strategic community to stay intellectually engaged with this issue so as to anticipate the increasingly apparent linkages between India and the Arctic.

BIBLIOGRAPHY

Borgerson, SG (2008). Arctic meltdown: The economic and security implications of global warming. *Foreign Affairs*, 87, 63–77.

Børresen, J (2008). The Arctic Highway. *The Polar Game: The Actors and Issue at Stake. Heartland-Eurasian Review of Geopolitics.*

Carey, S. *The UK's Role in Diego Garcia: Green Fingers or Red Faces?* Newstatesman, 07 September 2009.

Council, N (2004). *Mapping the Global Future: Report of the National Intelligence Council's 2020 Project.* Washington, DC: US Government Printing Office.

Council, A (2006). Arctic Marine Shipping Assessment: *The Arctic Council's Response to Changing Marine Access.* PAME, Progress Report.

Council, A (2008). *The Future of Arctic Marine Navigation in Mid-Century.* An Assessment Undertaken by the Protection of Arctic Marine Environment Working Group of the Arctic Council.

Drummond, KJ (2000). *Energy-Resources Map of the Circum-Pacific Region.* U.S. Geological Survey, scale 1:10,000,000.

IEA (2005). *Resource to Reserve: Oil & Gas Technologies for the Energy Markets of the Future.* International Energy Agency.

IPCC (2007). In *Climate Change 2007: Synthesis Report. Contribution of Working Groups I, II and III to the Fourth Assessment Report of the Intergovernmental Panel on Climate Change*, RK Pachauri and A Reisinger (eds.), pp. 104. Geneva, Switzerland: IPCC.

IPCC (2007). *Climate Change 2007: The Physical Science Basis. Contribution of Working Group I to the Fourth Assessment Report of the Intergovernmental Panel on Climate Change*, S Solomon, D Qin, M Manning, Z Chen, M Marquis, KB Averyt, M Tignor and HL Miller (eds.), pp. 996. Cambridge, United Kingdom and New York, NY, USA: Cambridge University Press.

Karas, TH (2003). *Global Climate Change and International Security.* The Sandia Report.

Krauss, C. Carving up the Arctic seabed — Two options: The "Meridian" method or the "Sector' solution" (10 October 2005). *The New York Times.*

Krysiek, SM (2007). The battle for the next energy frontier: The Russian polar expedition and the future of Arctic hydrocarbons. *Oxford Energy Comment.* London: Oxford Institute of Energy Studies.

Prestrud, P (2008). Climate change and polar region. *The World Ocean in Globalisation Challenges for Marine Region*, Oslo.

Richardson, M (2007). Quest for Arctic Reserve: The last Big Cold Rush Heats Up. *The Strait Times*, 31 August.

Theutenberg, BJ (2000). Development and cooperation in the Arctic. Chapter 14. In *The Arctic Challenge: Nordic and Canadian Approaches to Security and Cooperation in an Emerging International Region*, K. e. Möttölä, p. 335. London: Westview Press.

US Geological Survey. Circum-Arctic Resource Appraisal: Estimates of Undiscovered Oil and Gas North of the Arctic Circle. *Fact Sheet 2008–3049.* USGS.

Wade, R (2008). A warmer Arctic needs shipping rules. *Financial Times*, London, 16 January.

Young, OR (1985–1986). The age of the Arctic. *Foreign Policy*, 61, 160–179.

ENDNOTES

[1] Young, OR (1985–1986). The age of the Arctic. *Foreign Policy*, No. 61, 160–179.

[2] Carman, J. Economic and strategic implications of an ice free Arctic. http://usna.academia.edu [accessed on 18 August 2011].

[3] Borgerson, SG (2008). Arctic meltdown: The economic and security implications of global warming. *Foreign Affairs.*

[4] IPCC (2007). *Climate Change 2007: The Physical Science Basis. Contribution of Working Group I to the Fourth Assessment Report of the Intergovernmental Panel on Climate Change*, S Solomon, D Qin, M Manning, Z Chen, M Marquis, KB Averyt, M Tignor and HL Miller (eds.), p. 996. Cambridge, United Kingdom and New York, NY, USA: Cambridge University Press.

[5] Holland, MM, C Bitz and B Tremblay (2006). Future abrupt reductions in the summer Arctic Sea ice. *Geophysical Research Letters*, 33, L23503.

[6] International Northern Sea Route Programme. *Fridtjof Nansen Institute.* http://www.fni.no/projects/insrop.html [accessed on 19 August 2011]

[7] NSROP simulation analysis. www.fni.no/insrop

[8] Borgerson, S (2008). Arctic meltdown: The economic and security implications of global warming. *Foreign Affairs*, 87, 63–77.

[9] Theutenberg, BJ (1988). Development and Cooperation in the Arctic. (Chapter 14). In the *Arctic Challenge: Nordic and Canadian Approaches to Security and Cooperation in an Emerging International Region*, Möttölä, K (ed.), p. 335. London: Westview Press.

[10] Nicoll, A and J Delaney (eds.), (2007). India's energy insecurity. *IISS Strategic Comments.* 9 November. www.iiss.org.stratcom

[11] Storey, I (2006). China's 'Malacca' Dilemma. Jamestown *Foundation, China Brief*, 6(8), 11 April.

[12] *Ibid.*

[13] Sea level rise, understanding the past, improving projections for the future. *National Research Flagships: Wealth from Oceans* . http://www.cmar.csiro.au/sealevel/ [accessed on 18 August 2011].

[14] Sea level rise, understanding the past — improving projections for the future. *National Research Flagships: Wealth from Oceans.* http://www.cmar.csiro.au/sealevel/sl hist intro.html [accessed on 24 January 2010].

[15] *Ibid.*

[16] A post IPCC AR4 update on sea level rise. Antarctic Climate and Ecosystems Cooperative Research Center. http://www.cmar.csiro.au/sealevel/ [accessed on 16 August 2011].

[17] *Ibid.*

[18] Dasgupta, K (2009). Swallowed by the rising sea. *Hindustan Times*, 1 December, Saving Our Planet section.

[19] Oceans and sea level rise: Consequences of climate change on the oceans. Climate Institute. http://www.climate.org/topics/sea-level/ index.html [accessed on 24 January 2010].

[20] Pender, J. Climate change and displacement — community-led adaptation in Bangladesh. http://www.fmreview.org [accessed on 29 January 2010].

[21] *Ibid.*

[22] Maldives Geography. http://www.global-seafood-company.com/UK-Malediven. html [accessed on 16 August 2011].

[23] Carey, S. The UK's Role in Diego Garcia: Green Fingers or Red Faces? *Newstatesman*, 07 September 2009.

[24] Svalbard Treaty. http://en.wikipedia.org/wiki/Svalbard treaty [accessed on 16 August 2011].

[25] Among the few exceptions were Sweden's war with Russia (1590–1595) and the Kalmar war (1611–1613) between Sweden and Denmark that both were about Arctic trade with Russia.

[26] Alaska was purchased by USA from Russian empire in 1867. Norway was assigned sovereignty over Svalbad during the Paris treaty of 1920.

[27] Nowlan, L. Arctic legal regime for environmental protection. http://data.iucn.org/dbtw-wpd/edocs/ [accessed on 16 August 2011].

[28] Contents. *United Nations Convention on the Laws of the Sea.* http://www.un.org/Depts/los/convention agreements/texts [accessed on 19 August 2011].

[29] Press Releases and Statement, Arctic Summit Freezes Out peoples and Three nations. http://www.sikunews.com/News/International/Arctic-Summit-freezes-out-Arctic-peoples-and-three-nations-7352.

[30] CIA Opens Center on Climate Change and Security. *Press Releases and Statement.* https://www.cia.gov/news-information/press-releases-statements/center-on-climate-change-and-national-security.html

[31] *Ibid.*

[32] Spykman, NJ (1944). *The Geography of the Peace*, p. 41. New York: Harcourt, Brace.

7

THE ARCTIC AND JAPAN: ENERGY SECURITY AND CLIMATE SECURITY

Hiroshi Ohta

7.1. INTRODUCTION

The objectives of this chapter are to think about the implications of a warming Arctic for the Arctic region and the rest of the world, and to contemplate Japan's interests and potential role for conserving both biological and societal diversity in this region. The key conceptual framework is shaped by two concerns: energy security and environmental security. This chapter's perspective, in today's world, is the context of the "new energy crisis", which arises from two conflicting policy objectives, namely to supply energy to feed economic growth and, at the same time, reduce greenhouse gases (GHGs) substantially.

The two central concerns of this chapter can be recaptured in the following question: how to pursue development while conserving the environment and distinctive cultures in the Arctic region. These policy objectives are the core elements of sustainable development in the Arctic, which address how to manage economic development that will be actively pursued when the snow, ice and permafrost melt, when the sea routes are open during the summer, and when access to natural resources becomes much easier. However, it is not difficult to imagine that rapid economic growth may bring immense environmental degradation and societal clashes to this region. What will be the consequences of a warming Arctic for the Arctic region and the world? Will the economic benefits from development be sufficient to compensate for the loss of biodiversity and distinctive cultures?

Not only littoral states, but also non-littoral states become concerned with access to the Arctic and its resources. As a non-littoral state, Japan has applied to the Arctic Council for obtaining observer status. When this chapter was completed (August 2011), the Arctic policy was not yet fully developed apart from calling for open and safe navigation and access to natural resources in the Arctic. Since Japan almost totally lacks strategic energy sources, energy security is a matter of the utmost importance. Japan is also very vulnerable to natural disasters such as typhoons, earthquakes, and tsunamis. In fact, after the oil crises in the 1970s, one private study spelled out the concept of comprehensive security, addressing military and non-military security concerns, including energy and food security as well as countermeasures against earthquakes. Here I find a potential role that Japan can play in international governance of the Arctic as a shared resource region by addressing the issues of inter-linkages among energy security and climate security.

In the first section of this chapter, some efforts to broaden the scope of the concept of security are introduced with the aim of setting up a conceptual framework for this chapter. These efforts include Japan's concept of comprehensive security and "environmental security". The next section sheds light on the Arctic and Japan with the focus on the impacts of a warming Arctic on the region and the world as well as Japan's basic stance toward Arctic affairs. The following section is devoted to the Japanese concept of energy security and strategies as well as the gist of its climate policy. The final section describes recent developments in the concepts of climate security.

7.2. BROADER CONCEPTS OF SECURITY

The conventional notion of security that is generally shared is the narrow sense of security within the framework of the international society consisting of nation states, in which a sovereign state is bound to defend its political independence, territorial integrity, and its people. Thus, the conventional understanding of security can be seen as a concept focusing on a state and its government, which is concerned with safeguarding territory, maintaining political independence and protecting, through military and other necessary means, the people and assets within its territory from invasion by outside enemies.[1]

However, this kind of state-centric or politico/military-centric view of security is ill-suited to deal with transnational terrorism, massive population movements, infectious diseases such as HIV/AIDS, and threats to the earth's environment — the global issues that transcend the territorial boundaries of states. The collapse of the Cold War system that was based on the balance of fear between two superpowers has lessened the sense of imminent danger of a nuclear confrontation. As a result, other security concerns have become more conspicuous. In this vein, this section presents two perspectives of the broader concepts of security.

7.2.1. *Japan's concept of comprehensive security*

We can find one of the most interesting efforts to broaden the scope of the concept of security even during the Cold War era. A report on Japan's concept of comprehensive security includes one example of multi sector security concerns. The opening statement of this report reads: "Security means protecting people's life from various forms of threat."[2] In April 1979, the late Prime Minister Masayoshi Ohira requested that a task force led by Dr. Masamichi Inoki, former head of Japan's Defence Academy, carry out a study outlining a comprehensive national security plan for Japan. In July 1980, shortly after Ohira's unexpected death, the Report of the Concept on Comprehensive Security was submitted to Acting Prime Minister Masayoshi Ito. It contains a conceptual definition of comprehensive national security, a discussion of the international situation and tasks to be faced, and analyses of specific issues.

Efforts required for addressing security concerns need to take place on three levels: turning the overall international environment into a favourable one; coping with threats autonomously; and as an intermediary measure, creating a favourable international environment within a limited scope while promoting solidarity with countries sharing the same ideals and interests.

The specific issues include Japan–US relations, defence capability, relations with China and the former Soviet Union, energy and food security concerns, countermeasures for large-scale earthquakes, and a discussion of crisis management.[3] Although the 1980 report introduced a concept of comprehensive security that remains within the traditional security framework, its authors were compelled to broaden their definition of security to include non-military concerns. The report addresses not only the military, but also the political and economic aspects of security. It stops short of referring to the social and environmental factors, but it does address concerns about natural disasters. It urges, for instance, comprehensive countermeasures against large-scale earthquakes, such as improving prediction techniques, as well as strengthening emergency management capabilities at the national and local government levels.

In retrospect, if this policy guideline for the preparation and management of major earthquakes were embodied as a systematic policy in Japan, the number of victims of the 1995 Hanshin-Awaji Earthquake might have been much smaller than the actual loss of over 6,400 lives. Furthermore, an estimated death toll of 200,000 from Haiti's earthquake of January 2010 warns us of the danger of major earthquakes to unprepared and defenceless communities. However, Japan's earthquake and tsunami of 11 March 2011 tell us that the countermeasures against earthquakes and tsunamis, based on the past records of disasters are not enough to protect people's lives. This makes us take a closer look at "environmental security."

7.2.2. *Redefining the concept of security: "environmental security"*[4]

There are three distinctive views of the relationship between ecology and security: traditional, alternative, and eclectic. Traditional views are represented by the dominant social paradigm of the realist perspective and the idea of a politics of scarcity.[5] According to the theory of the politics of scarcity, scarcity may develop into international conflict when there is restricted access to vital non-renewable and renewable natural resources such as oil, coal, water as well as food. Thomas Homer-Dixon represents this school of thought.[6] Other prominent environmentally concerned theorists and policy advocates who base their arguments on this concept of ecological security are Lester Brown and Jessica Mathews.[7] They stress that global environmental threats or social and/or political disorder can be caused by "environmental refugees" moving across national borders. Richard Ullman extends this concept of security to include natural disasters including earthquakes.[8]

The alternative view of environmental security is framed from the ecological perspective and is based on Harold and Margaret Sprout's pioneering works in international relations.[9] The ecological perspective envisages "international politics as a *system of relationships* among *interdependent*, *earth-related* communities that share with one another an *increasingly crowded planet* that offers *finite* and *exhaustible* quantities of *basic essentials* of human well-being and existence" (emphasis in original).[10]

As to the root cause of ecological crisis, this perspective points to "a transformation both in human capacity to cope with nature and in human attitudes toward nature and other men."[11] In other words, rapid technological advancement and the "narrower engineering perspective" drive human beings to further exploit natural resources and reshape nature. The "engineering perspective" is that which results in the invention of new technologies without concern for side effects which damage human health or the natural environment.[12] Similarly, according to the most current advocates of the ecological perspective, the prevailing social perspective reflects an industrial culture that shapes the structure of social institutions. In other words, "there is a general set of values, attitudes, beliefs, and perceptions that are shared by most members of industrial societies."[13] The proponents of this view of the ecology–security relationship argue that environmental security will demand changes in attitudes, beliefs, and perceptions if an environmentally secure world is to be realized.

The eclectic view, finally, revolves around the concept of "sustainable development," as articulated in the 1987 report of the World Commission on Environment and Development (also known as the Brundtland Commission after the name of the chairperson of this Commission).[14] The Commission called for radically different approaches to economic development, equity, resource management, and related issues in order to meet the challenge of a rapidly growing world population. Although it has acknowledged the necessity of development in order to break the vicious cycles linking poverty to population growth and environmental destruction, it has rejected unlimited development. The Commission defined sustainable development as meeting "the needs of the present without compromising the ability of future generations to meet their own needs."[15]

Some caveats, however, should be mentioned regarding the notion of environmental security as an analytical concept in international relations. Although both military violence and environmental disasters kill people and destroy property, we do not normally regard an earthquake disaster or a destructive hurricane as a national security issue. Violence perpetrated by humans is highly intentional: "organizations are mobilized, weapons procured, and wars waged with relatively definite aims in mind, whereas environmental degradation is largely unintentional, the side effect of many other activities."[16] Moreover, while people and their property may be protected from external threat by a highly centralized and hierarchical military organization, the environment can only be protected by the participation of everyone involved in preserving fragile ecosystems at national, regional, and global levels. The sceptics thus argue: "If everything that causes decline in human well-being is labelled a security threat, the term loses any analytical usefulness. Because national security from violence and environmental habitability has little in common, the new fashion of linking them may create a conceptual muddle rather than a paradigm shift."[17]

We should be attentive to these caveats, but the consequences of global warming expose the peoples of small island states and states with long coastal areas to imminent danger to their survival due to rising sea levels. In this case, a general term like "environmental security" or a more particular term like "climate security" becomes useful in prioritizing the measures that people and their communities can use to cope with environmental threats.

7.3. THE ARCTIC AND JAPAN

7.3.1. *The Arctic's geopolitics: resources, peoples and the environment*

Unlike the Antarctic where there is no serious conflict over territorial claims among concerned nations, the Arctic has had geopolitical dynamisms over territories and resources. The concerned land areas of the Arctic include Alaska (except for the area known as Southeast); the Yukon and Northwest Territories, Northern Quebec, and all of Labrador in Canada; all of Greenland; Iceland; the Northern countries of Norway, Sweden, and Finland (known collectively as Fennoscandia), as well as all of what the Russians treat as the Arctic and the Russian North. The marine systems of the Arctic Ocean and its adjacent seas are the Bering, Chukchi, Beaufort, Greenland, Norwegian, Barents, Kara, Laptev, and East Siberian Seas.[18]

Although international society tends to see the Arctic through the lenses of geopolitics and a new economic frontier, the development of the Arctic Council provides us with a model or a good testing case for achieving sustainable development and global governance. Sustainable development requires meeting three major needs of human society without depriving the future generations of their ability to meet the same needs: economic, social and environmental.[19] Sustainable development in the context of the Arctic is succinctly stated by O. Young: That is, "the establishment of economic systems capable of maintaining themselves over time without disrupting major Arctic ecosystems or destroying the distinctive cultures of the Arctic's permanent residents."[20]

In order to establish a sustainable society in the Arctic region, both state and non-state actors have to participate in managing natural resources while maintaining both biological diversity and social diversity. The governing body of the Arctic Council[21] has a potential to achieve the goal of sustainable development through a scheme of global governance in which non-state actors are considered as important as state actors. While eight Arctic States: Canada, Denmark, including Greenland, Finland, Iceland, Norway, Russia, Sweden, and the United States have to cooperate with each other, they should also allow the substantive participation of the concerned non-state actors in regional governance. So, the Permanent Participants of the Arctic Council are crucial for maintaining the integrity of the Arctic region. Those important non-state actors include the *Aleut International Association* (AIA), *Arctic Athabaskan Council* (AAC), *Gwich'in Council International* (GCI), *Inuit Circumpolar Council* (ICC), *Russian Association of Indigenous Peoples of the North* (Raipon), and *Saami Council*.[22] In addition, scientific communities, various environmental and human rights groups as well as industry/business interests can also contribute to the Arctic affairs.

Having set the basic framework for how to perceive international relations in the Arctic as a model for sustainable development and global governance, the current state of the Arctic environment, which is changing rapidly due to global warming, is discussed.

7.3.2. The main findings of the Arctic climate impact assessment

The Arctic Climate Impact Assessment (ACIA) was an independent scientific study carried out by hundreds of scientists over four years and its results were published in 2004. The ACIA evaluates the Arctic climate change and its impacts for the region and the world. This assessment was called for by the Arctic Council and the International Arctic Science Committee (IASC), and charged the Arctic Monitoring and Assessment Program (AMAP) and the Conservation of Arctic Flora and Fauna (CAFF) with its implementation.[23] Here are some of ACIA's major findings.

One of the findings is that temperatures in the Arctic are now rapidly rising and much larger changes are still projected.[24] In most places, temperatures in winter are rising more rapidly than in summer. In Alaska and Western Canada, for instance, winter temperatures have increased as much as 3–4°C (5–7°F) over the past 50 years. Declining sea ice is also remarkable. Over the past 30 years, the annual average sea ice extent has decreased by about 80%, or 617,760 square kilometres (386,100 square miles), an area larger than all of Norway, Sweden, and Denmark (or Texas and Arizona) combined, and the melting trend is accelerating. Loss of sea ice during the summer is projected to be considerably greater, with some models showing near-complete disappearance of summer sea ice.[25]

Another major finding is the global impact of Arctic warming. The primary factors for sea level rise are thermal expansion due to global warming and melting of land-based ice. Land ice in the Arctic is concentrated in the Greenland Ice Sheet. Though there are considerable variations from year to year, the largest surface-melt area on the ice sheet of Greenland increased on average by 16% from 1979 to 2002, which was roughly as large as the size of Sweden. If the Greenland Ice sheet were to completely melt, it would result in a sea level rise of about 7 meters. Changes in the Arctic can impact global climate in two other ways. As snow and ice melt and vegetation cover changes, the reflectivity of the land and ocean surface changes. When the surface becomes darker, it absorbs more of the sun's energy, amplifying global warming. Another mechanism involves potential changes to the global ocean circulation (also known as the conveyor belt circulation) that carries warmth from the tropic regions to the Northern Hemisphere, including Western Europe and the Scandinavian region. This global ocean circulation is driven by the salinity balance, which is now being disturbed by more fresh water pouring into the oceans through melting Arctic ice and river runoff.[26]

Warming and climate change in the Arctic region also affect animal species' diversity, ranging from their habitats to their distribution. The most popular example is polar bears, which are heavily dependent on sea ice for hunting seals and for corridors to move from one area to another. If sea ice cover completely disappears during the summer, polar bears cannot survive and this has significant consequences for the ecosystem in this region. Likewise, ice-dependent seals, which give birth to and nurse their pups on the ice, are being threatened by the melting of the sea ice. They include the ringed seal, ribbon seal, and bearded seal. Arctic warming further affects migratory birds, caribou and reindeer. Several hundred million birds migrate to the Arctic each summer and breed and nest on tundra with a swarm of their insect food sources. Similarly, caribou and reindeer depend on abundant tundra vegetation and the wide range of foraging grounds. While a warmer climate in this region allows the tree line to advance northward encroaching on tundra, it creates unfavourable conditions for migratory birds and caribou and reindeer.[27]

ACIA's findings also point out economic and cultural impacts on indigenous communities, coastal communities and their facilities. The decrease in the populations of polar bears, seals, caribou and other animals is the major economic and social threat to indigenous peoples like Inuit. It, therefore, holds: "Changes in species' ranges and availability, access to these species, a perceived reduction in weather predictability, and travel safety in changing ice and weather conditions are making people feel like strangers in their own land."[28] Moreover, coastal communities and their facilities face a growing problem of coastal erosion caused by the sea level rise and a reduction of sea ice. Even inland areas are affected by Arctic warming due to serious consequences of thawing permafrost, namely, disruption of transportation, buildings, and infrastructure. As permafrost on land thaws, buildings, roads, pipelines, and industrial facilities are likely to be damaged or destabilized.[29]

Moreover, the ACIA suggests that climate change and other influences interact to affect human health. Although such impacts differ from region to region and also depend on the presence or absence of communities' adaptability or health service availability, rural Arctic residents in small and isolated communities tend to suffer most from climatic change and ensuing environmental degradation, especially those communities relying on subsistence hunting and fishing of a few species.[30]

Among these findings, the prospects for the increase of marine transport and access to resources in this region involve economic interests, geopolitical and environmental concerns. Some model projections point out that sea ice in summer will retreat away from most Arctic landmasses so as to enable opening of the Northern Sea Route through the Russian North to the Bering Sea or the Northwest Passage through Northern Canada and Alaska to the Bering Sea.

These trade routes between Europe and Asia are much shorter and quicker than the routes through the Panama Canal or Suez Canal and have tremendous economic potentials for both regions. However, the possibility of opening new sea routes raises questions about security and safety, as well as environmental concerns.[31] In addition, the melting ice in the Arctic will increase access to seabed resources. However, offshore development of resources has to be carried out with utmost care and attention. We have to remind ourselves of the oil spill of the Exxon Valdes in Alaska in 1989 and the explosion of BP's offshore oil rig in the Gulf of Mexico in April 2010.

These are the summaries of the impacts of global warming on the Arctic and the world. The melting ice in this region may open up new economic opportunities, but at the same time poses threats to ecosystems and local communities.

7.3.3. *The Arctic and Japan*

In 2008, when Mr Koji Tsuruoka, the Director-General of the International Legal Affairs of the Bureau of the Ministry of Foreign Affairs (MOFA), was asked why Japan did not apply for observer status to the Arctic Council, he was quoted as saying, "We cannot do so without an adequate budget. It is a matter of policy priority."[32] However, six months later, the Japanese government announced its plan to start application procedures to the Arctic Council for observer status at the 50th Anniversary of the Antarctic Treaty held in Washington, D.C. on 6 April 2009.[33]

The main reasons are environmental, strategic, and economic.[34] With regard to the Arctic, the Japanese government has been most concerned about the decreasing ice-covered areas due to global warming and the ensuing drastic climatic change in this region. What is happening there is a sign of what is going to happen across the globe, including the mid-latitude areas. The government also has concerns about the fragility of the Arctic environment, where harmful persistent organic pollutants (POPs) are stagnant or circulate within the Arctic Ocean as they lose mobility in lower temperatures.

Strategic concerns are intertwined with economic interests. The thawing of ice in the summer opens up a new navigation route. For instance, the navigation route between Yokohama and Hamburg via the Arctic would be 38% shorter than the route via the Suez Canal.[35]

Transportation and logistics industries have great interest in the Arctic route. Thus, the Japanese government wants to ensure the open and safe navigation in the Arctic. The decrease of ice will also facilitate development of natural resources in this region, which is rich in strategic natural resources such as oil, natural gas, lead, zinc, and rare earth.

Having these environmental, strategic and economic concerns, the Japanese government considers the Arctic the common heritage of mankind to be used for peaceful purposes. It states that non-Arctic states should have access to the natural resources in the Arctic and, at the same time, protect its environment so as to pursue sustainable development. Above all, Japan is concerned about and is continuously making efforts to mitigate climate change. The Ocean Division of the MOFA's International Legal Affairs Bureau is in charge of Arctic issues and collaborates with other relevant ministries and agencies, scholars, NGOs as well as the private sector.[36] On 2 September 2010, the MOFA established the Arctic Task Force (ATF) "to make cross sectoral approach towards the foreign policy on the Arctic"[37] and applied for observer status in the Arctic Council. At the time of writing, Japan had yet to obtain observer status.

7.4. THE "NEW ENERGY CRISIS," ENERGY SECURITY AND JAPAN'S STRATEGY

7.4.1. *The new energy crisis*

The energy crisis we face today is different from the two major oil crises of 1973–1974 and 1979–1980 when the lack of oil from the Middle East was the primary cause. Today's energy crisis stems from two conflicting policy goals: how to supply energy for economic growth and, at the same time, reduce GHGs substantially. One energy expert calls this conundrum "the equation of Johannesburg", where the World Summit on Sustainable Development was held in 2002.[38]

What does this "equation of Johannesburg" mean in more concrete terms and how can we solve it? The International Energy Agency (IEA) developed three scenarios for projecting different paths to the future up to 2030: a reference scenario (or business as usual, or BAU), an alternative policy scenario with some significant policy changes, and a 450 ppm stabilization scenario.[39] In 2005, global CO_2 emissions amounted to 27 gigatons (Gt) while according to IEA's reference (or BAU) scenario, they will reach 42 Gt in 2030. If we want to avoid an increase in the global mean temperature of more than 2°C from the level of the pre-Industrial Revolution, we have to stabilize the concentration of CO_2 in the atmosphere at the level of 450 ppm. To meet this goal, however, we should reduce the emissions of CO_2 to 23 Gt by 2030, a substantial reduction of 19 Gt from that of the BAU scenario. What means are available to us or need to be developed? The IEA suggests that we develop the technology of carbon capture and sequestration (CCS), build more nuclear power plants, utilize more renewable energy sources, improve energy efficiency, and conserve energy.[40]

However, the nuclear option is not viable for Japan, especially after the nuclear catastrophe of 11 March 2011(discussed later). Instead, Japan must further rely on fossil fuels until renewable energy sources can become the primary energy sources running the Japanese economy. Regarding the reduction of GHG emissions, or the other element of the "Johannesburg equation", Japan is likely to expand the supply of natural gas, which emits less CO_2 than either coal or oil does. In this context, it can be expected that Japan would become interested in investing in the development of natural gas in such regions as the Russian Arctic and Norway whenever such an opportunity arises. In addition, since the diversification of primary energy sources has been the core of the Japanese energy policy since the first oil crisis, the Arctic energy resources appear very attractive to both the public and private sectors in Japan. Therefore, it is important for us to understand the heart of Japanese energy concerns in order to draw some kind of inference about Japan's involvement in the development of the Arctic region in the foreseeable future.

7.4.2. *Japan's concept of energy security and energy strategy*

Japan, a poor country in terms of non-renewable natural resources such as oil and natural gas, but rich in industrial activities, also faces this new energy crisis. However, an additional feature to Japan's "equation of Johannesburg" is its heavy dependence on imported fossil fuels, particularly from the Middle Eastern countries. Since the two oil crises of the 1970s, the Japanese government's strategic concern has been to diversify the energy sources and reduce dependence on the Middle Eastern sources. Thus, this section addresses the structural problems relating to Japanese energy supply and demand, and aims to provide a backdrop for Japan's quest for future access to Arctic natural resources.

7.4.2.1. *The context of energy security discourse in Japan*

Japan's energy self-sufficiency rate is merely 16%, meaning it must import most of its energy requirements to drive its economy. Japan is the largest importer of liquefied natural gas (LNG) and coal, and the third largest importer of crude oil, after the United States and China. The lack of domestic hydrocarbon resources pushes Japanese energy companies to take part in upstream foreign oil and gas developmental projects while Japanese manufactures export energy-sector capital equipment to the world.[41]

In 2010, total energy consumption in Japan amounted to 500.9 million tons oil equivalent (mtoe). Of this, the largest source of energy is oil (201.6 million tons), followed by coal (123.7 mtoe), natural gas (85.1 mtoe), nuclear energy (66.2 mtoe), hydro-electricity (19.3 mtoe), and renewables (5.1 mtoe).[42] However, oil's share of total energy consumption declined to about 40% in 2010 from about 80% in the 1970s.[43] In passing, coal's share of total energy consumption (as of 2010) was about 25%, natural gas about 17%, and nuclear energy about 13%. The shares of hydropower and other renewable sources were relatively small, at about 4% and 1%, respectively. (See Fig. 7.1)

Since Japan's experiences of the two oil crises in the early and late 1970s were extremely severe and the fundamental vulnerability of its dependence on outside energy sources has been unchanged, it is quite natural that the discourse on energy in Japan has been dominated by concerns about energy security. The following graph shows (See Fig. 7.2) the background of such concerns. Despite its efforts

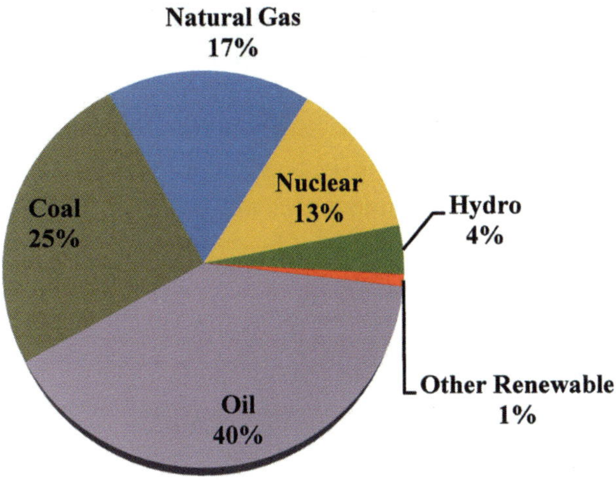

Fig. 7.1. Japan's Total Energy Consumption, 2010.
Source: Created by the author based on the data available in *BP Statistical Review of World Energy June 2011.*

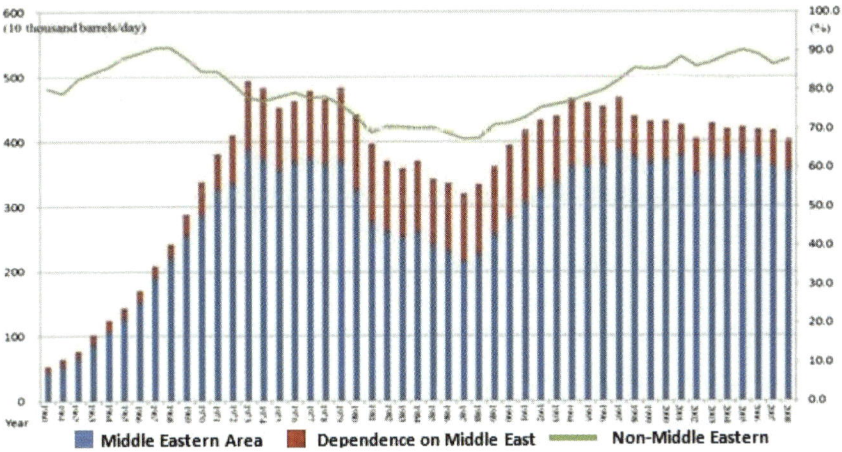

Fig. 7.2. Japan's Oil Imports and Dependence on the Middle East: 1961–2008.
Source: Created by the author based on the data provided by: The Energy Agency of the METI, *Enerugi hakusho 2010 (The White Paper of Energy 2010)*. (Available at http://www.enecho.meti.go.jp/topics/hakusho/2010energyhtml/index.html [accessed 28 August 2011].)

to diversify energy sources after the two oil crises, Japan has been heavily dependent on oil as Japan's main primary energy source. Above all, the fact that Japan has been heavily dependent on Middle Eastern oil is another root cause of its serious energy security concerns, especially in the present situation and the recent hike in the price of crude oil, and increasing world energy demands.

Therefore, among many policies for reducing Japan's dependence on foreign energy sources, both the public and private sectors have steadily increased their stockpiles of oil to 5–6 months' cover, along with energy conservation efforts.

7.4.2.2. *METI's concept of energy security*

It is quite difficult, if not impossible, to find a policy guideline that spells out Japan's concept of energy security. However, if there is anything close to it, an interim report of the Energy Security Study Group of the Ministry of Economics, Trade and Industry (METI) of Japan is certainly one of them. This *Interim Report* released in 2006 discusses the objective of energy security, the major energy risks, energy security policy and measures, and a specific action plan. A look at its synopsis is worthwhile.

The objective of energy security is two-fold according to the report.[44] The two objectives are: to guarantee enough energy necessary for "the domestic economy and livelihood at an appropriate price" and to seek coexistence with the world economy by "working with the international community to address and to contribute to solving Asia's and the world's energy problems."[45]

Within this context, the Interim Report identifies five major energy risks, including (1) political conditions in the Middle East; (2) terrorism, natural disasters, and accidents (misconducts); (3) reduction of investments in supply nations (oil and/or gas exporting countries); (4) trends of demand nations (China, India, etc.); and (5) issues facing the energy industry.[46] As the first two major energy risks are not relevant to energy security issues relating to the Arctic region, only the latter three risk concerns are summarized in the following paragraphs.

The major energy risk relating to energy supplying nations stems from governmental intrusion into the management of energy industries or nationalization of energy industries so as to restrict foreign investment in their energy industries. A country like Russia considers its energy resources strategic in terms of exerting influence on importing nations. Increased risks in major oil and gas-producing nations and growing government involvement in the energy sector ("politicization" of energy markets) have made investment conditions worse for foreign investors, who can invest in plants and equipment and bring advanced technologies for more efficient production. Without sufficient foreign investment there will be a medium- to long-term decline in the supply capacity of major energy exporting countries. Thus, governmental regulations in major energy supplying countries can constitute a structural risk to energy security.[47]

On the demand side, the rapid economic growth of emerging economies such as China and India could cause shortages of energy and result in medium- to long-term strains in the international energy market, according to the *Interim Report*.[48] Moreover, as these countries face increasing dependency on foreign energy sources, they might adopt aggressive policies for securing their energy sources. In the worst case scenario, they might take exclusionist measures to obtain vital energy necessary for maintaining their economic growth.

Finally, the *Interim Report* points out the risk for Japan's domestic energy industries. Accordingly, the deregulation of the Japanese energy markets negatively affects Japan's energy industries.[49] The rationale for such deregulation is to introduce competition so as to reduce energy costs while increasing efficiency. However, the report argues that this deregulation results in uncertainty about the future economic environment and the Japanese energy industry's ability to secure demand.[50] Furthermore, the population decline in Japan will likely lead to a decrease in energy demand and, in turn, to the reduction of investment in plant and equipment as well as research and development for new technologies. As a result of these structural changes, the supply capacity of the Japanese energy industries is likely to decline posing a threat to Japan's energy security.[51]

Having identified major risks, the report suggests some policies and measures to overcome these risks. These can be divided into two: international and domestic. The foci of the international policies and measures includes strengthening co-operation with Asian countries

(China, India, and others), consolidating the relationship with energy supply nations, and strengthening international and regional frameworks for co-operation such as G8 summits, IEA, ASEAN+3, and APEC. It also includes consolidating the relationship with major powers, namely the United States, China, India, and Russia. The domestic policies and measures are those which help enhance Japan's response measures, improve management systems, fortify the energy supply and demand structure, improve energy response measures and increase oil reserve capacity, as well as emergency response measures for natural gas.[52]

The Interim Report also sheds light on more concrete action plans with regard to co-operation with the Asian countries. Above all, co-operation in the areas of energy conservation, promotion of oil reserves, risk management partnerships and environmental protection may have promising future prospects for ensuring energy security not only for Japan, but also Asia as a whole. However, the proposed cooperation with Asian countries in the area of nuclear energy use needs a major review following the Fukushima Nuclear Power Plant incident in March 2011.

7.4.2.3. *New national energy strategy*

One month prior to METI's Interim Report mentioned above, another basic energy policy paper was issued by METI in conjunction with Japan's climate change policy. *The New National Energy Strategy* set up very high and ambitious targets.[53] Thus, regarding the energy conservation target, a 30% improvement in energy efficiency should be attained by 2030. Japan's oil dependence as its primary energy source should be reduced to less than 40% by 2030. Above all, the transportation sector would be asked to reduce oil dependency by 80% by 2030. Meanwhile, overseas natural resources development would be encouraged so as to ensure the availability by Japanese developers of 40% of total oil imports by 2030. Nuclear power generation should be increased by 30% to 40% or more by around 2030.[54] Despite future plans for the expansion of new energy technologies, the role assigned them was very limited as the report failed to mention them as one of the components of the Japanese energy strategy.

However, the June 2010 cabinet meeting adopted the Basic Energy Plan. According to this, 70% of the electricity would be generated from non-fossil fuels or "zero emissions" by 2030 of which the share of renewable energy and nuclear energy would be 20% and 50%, respectively.[55] Towards this end, five additional nuclear power plants would be built by 2020 and another nine by 2030.[56]

Again, this energy strategy has to be fundamentally revised following the Fukushima Dai-ichi (No. 1) Nuclear Power Plant incident which occurred on 11 March 2011, when a 9.0 magnitude earthquake struck off the coast of Sendai, Japan, and a huge tsunami swept away and destroyed important facilities and equipment, including the emergency core cooling system (ECCS). The damage caused by the earthquake and ensuing tsunami forced Tokyo Electric Power Co. (TEPCO) to decommission four nuclear reactors (No. 1 to No. 4 reactor) of the Fukushima Dai-ichi nuclear complex. A high level of radiation is still being detected in Fukushima Prefecture.[57] According to the Nuclear and Industrial Safety Agency (NISA), this incident was classified as level seven, that which was the same as for the Chernobyl incident that occurred in 1986.[58] The sense of distrust towards nuclear safety is now rapidly spreading among the Japanese along with the shared sense of "risk society" à la Ulrich Beck.[59] In Japan, there are 54 operating nuclear reactors with a total installed generating capacity of about 49 GW.[60] Two reactors are now under construction and another 12 are in planning stages. However, few Japanese think this plan is feasible.

7.4.2.4. *Basic climate change policy*

According to the government's "Guidelines for Measures to Prevent Global Warming," Japan is committed to reducing GHGs by 6% below the 1990 level during the first commitment period of 2008–2012 under the Kyoto Protocol. The Guideline divides this objective into four categories.[61] The first is to reduce CO_2, CH_4, and N_2O by 2.5%. The second is to reduce CO_2 by 3.9% through sequestration via afforestation and reforestation under Kyoto Protocol Article 3.3 and through additional activities such as forest management under

Protocol Article 3.4. The third category actually allows increases of HFC, PFC, and SF6 by 2%. The remaining reductions will be achieved through utilization of the Kyoto mechanism.

As for the post-Kyoto commitment for a mid- and long-term reduction target, the cabinet of the former Prime Minister Yukio Hatoyama approved on 12 March 2010 the Basic Global Warming Bill.[62] This bill set a mid-term target for 2020, to reduce GHG emissions by 25% below the level of 1990, on the premise of "the establishment of a fair and effective international framework in which all major economies participate and on the agreement on ambitious targets by all major economies." The basic policy proposals to achieve this goal include (1) establishment of a domestic emissions trading system; (2) implementation of global a warming tax in the fiscal year 2011; (3) introduction of the system to purchase all electricity generated by renewable energy sources at a fixed price (Japan's feed-in-tariff system); (4) nuclear energy promotion with due consideration to safety; (5) further improvement in energy efficiency; and (6) promotion of R&D for innovative technologies. However, this bill was not passed during the ordinary Diet (Japan's Parliament) session in June 2010 and was dropped.

Given the uncertainty of Japan's domestic politics and the absence of a visionary political leadership,[63] there is no prospect in today's Japan for departing substantially from the previous energy policy. Thus, it is likely that Japan will become an active participant in the development in the Arctic energy development in the future. However, since the vast majority of the Japanese public prefers the use of renewable energy resources instead of fossil fuels and nuclear energy, there is a possibility for change while addressing the problem of climate change.[64]

7.5. THE CONCEPT OF CLIMATE SECURITY

Solving the "equation of Johannesburg" is quite difficult not only for Japan, but also for the rest of the world. Not a single national government can survive without advocating economic growth, which has been driven mainly by fossil fuels. However, as the impact of climate change on the Arctic is increasingly becoming a loud warning alarm signalling the destruction of the environment, (see Sec. 7.2 of this chapter), the concept of climate security has begun to attract a great deal of attention. When we attach the term "security" to climate change and if such designation is accepted by the local, national, or international communities, the issue takes politics beyond the established rules to be framed as a special kind of politics or as above politics.[65] It means that climate change "is presented as an existential threat, requiring emergency measures and justifying actions outside the normal bounds of political procedure."[66] Here we may be able to find the solution for the "Johannesburg equation" by placing the strong warning signals emanating from the Arctic region into the conceptual framework of climate security.

The term climate security has begun to be used at international fora, in the diplomatic world and policy-making circles, at least, in the developed countries. The former UN Secretary General Kofi Annan's address to the Climate Change Conference in Nairobi, in November 2006, referred to global climate change as "an all-encompassing threat".[67] This threat includes threats to health, food supply, the foundation of life (threat of sea level rising and natural disasters), peace and security. He further maintains: "Global climate change must take its place alongside those threats — conflict, poverty, the proliferation of deadly weapons ... "[68] Similarly, the current UN Secretary General Ban Ki-moon regards climate change as not merely an environmental problem, but a serious social and economic issue. He also notes that climate change, along with energy scarcity, has "implications for peace and security".[69]

At the First Session of the 110th Congress of the United States, the Biden–Lugar Resolution called on the government to address global climate change by returning to international negotiations. This resolution recognizes that the potential impacts of global climate change, such as "long-term drought, famine, mass migration, and abrupt climatic shifts", may generate international conflicts closely related to US national interests.[70] Likewise, in September 2006, Margaret Becket, then British Secretary of State for Foreign Affairs, gave a speech to the American business community on "Climate

Security: Risks and Opportunities for the Global Economy".[71] After identifying several impacts of climate change, she maintained: "climate insecurity means greater global insecurity." Then, she urged the American business community to rise to this challenge by saying: "It (climate insecurity) makes managing commercial risk much harder. And it means tackling global warming is a business imperative, not a business choice. And the second equally basic commercial imperative for action: staying competitive."[72]

The EU's paper from the High Representative and the European Commission to the European Council entitled, "Climate Change and International Security", also addresses the issue of security threats from climate change. This report lists seven major concerns relating to climate change. They include: (1) conflict over resources; (2) economic damage and risk to coastal cities and critical infrastructure; (3) loss of territory and border disputes; (4) environmentally-induced migration; (5) situations of fragility and radicalization; (6) tension over energy supply; and (7) pressure on international governance.[73] This report also takes a comprehensive approach to address security concerns about the consequences of climate change.

In addition, a policy study carried out by Japan's Ministry of the Environment attempts to conceive climate change as a security concern.[74] According to this study, entitled *Report on Climate Security* (hereafter the *Report*), the threats from climate change are all-encompassing, including threats to food production, public health, economic activities and ecosystems. The economic and social damage caused by abnormal weather events is becoming increasingly more destructive than before and are hitting poor and vulnerable people and their communities hard. The sea level rise adds extra stress to these people and communities. In the worst case scenario, the sea level rise compounded with other forms of environmental degradation may force people to abandon their communities and migrate to some other places as "environmental refugees." The Report also suggests that the sea level rise might cause some territorial disputes, for example, due to the alteration of EEZs through submergence of territorial islands or the erosion of the seashore. Therefore, the *Report* argues, the impacts of the various threats of climate change "can be expected to build on one another, potentially leading to conflicts between nations."[75]

As for the usefulness of the concept of climate security, according to the *Report*, "framing the climate change issue as a security-related concern raises the political priority placed on the issue both domestically and internationally."[76] If nations, the international community, corporations and citizens all heed the growing and imminent threats of climate change, the policies and measures both to mitigate and adapt to it will be given higher policy priority both domestically and internationally. The policies and measures for reducing greenhouse gas emissions must be taken globally so as to begin establishing a low-carbon society.[77] The concept of climate security may also help promote international technological and financial transfer from developed countries to developing countries. Moreover, the recognition of climate security might generate a sense of international solidarity and promote international actions to assist "adaptation measures" for low-lying and island nations and other countries vulnerable to the consequences of climate change such as rising sea levels. Finally, the Report suggests that widespread awareness of climate security may exert pressure on the large emitting nations to actively join international efforts to stabilize the global climate system.[78]

In short, the impacts of climate change on the Arctic have already shown the international community what the rest of the world should expect in the foreseeable future if they do not take any immediate action to substantially reduce GHG emissions. It is hoped that the concept of climate security could "securitize" the issue of climate change so that we can properly draw the lessons from what the people in the Arctic are already experiencing as a result of climate change.

7.6. CONCLUSIONS

According to Young, the establishment of "economic systems that are capable of maintaining themselves over time without disrupting major Arctic ecosystems or destroying the distinctive cultures of the Arctic's permanent residents"[79] is crucial to governing the Arctic, a shared-resource region. As one of the non-littoral states, Japan wants free and safe navigation, as well as access to natural resources in the Arctic. The resources in this region will become important for the diversification of Japan's energy sources in the future. In the meantime, Japan can contribute to this region by helping realize sustainable development, above all conservation of biological diversity and cultural diversity.

Energy security is a matter of the utmost importance for Japan since it imports most of its energy requirements. Since the two oil crises of the 1970s, Japan has attempted to become less dependent on fossil fuels, especially oil. It has reduced the share of oil as one of its primary energy sources from more than 70% in the 1970s to less than 50% in the 2000s. However, dependence on Middle Eastern oil has increased again since the late 1980s. The unstable political and social situations in the Middle Eastern countries are one of the main causes of energy insecurity in Japan. One of Japan's strategies for overcoming this problem has been the nuclear option. Yet, the 11 March earthquake, tsunami, and ongoing paralysis of the Fukushima Diichi nuclear power plant appear to remove the nuclear option from Japan's energy security strategy.

It is now time for us to remind ourselves of the opening statement of the report about the concept of comprehensive security: "Security means protecting people's lives from various forms of threat." Based on this principle, I think Japan can play a substantial role in international governance of the Arctic as a shared resource region by becoming a champion of renewable energy, and a nation of energy conservation experts and community-based international co-operation professionals. Japan's unpopular and beleaguered prime minister, Naoto Kan officially announced his resignation on 26 August 2011, when a bill to promote the use of renewable energy by introducing feed-in tariffs was passed at the Japanese Diet. This legislation opens up the possibility for Japan to become less dependent on both fossil fuels and nuclear energy and thus drive the Japanese economy, at least, without relying on the Arctic's fossil fuels in the future. Whether or not we succeed in navigating Arctic affairs in a direction towards sustainable development could determine human society's destination.

BIBLIOGRAPHY

The Arctic Climate Impact Assessment (ACIA) (2004). *Impacts of A Warming Arctic.* Cambridge: Cambridge University Press.

Beck, U (1992). *Risk Society: Towards New Modernity.* London: Sage.

Barnett, RW (1984). *Beyond War: Japan's Concept of Comprehensive National Security.* Washington, D.C.: Pergamon-Brassey's.

Brown, LR *et al.* (1986). *State of World Environment.* New York: Norton.

Buzan, B, O Wæver and J deWilde (1998). *Security: A New Framework for Analysis.* Boulder, CO: Lynne Rienner.

Carr, EH (1939). *The Twenty-Years' Crisis, 1919–1939: An Introduction to the Study of International Relations.* London: Macmillan.

Chevalier, J-M (ed.) (2009). *The New Energy Crisis: Climate, Economics and Geopolitics.* New York: Palgrave Macmillan.

Deudney, D (1995). Environment and security: Muddled thinking. In *The Global Agenda: Issues and Perspectives*, CW Kegley, Jr and ER Wittkopf (1995), 4th Ed, 446–454. New York: McGraw-Hill.

Dyer, HC (1996). Environmental security as a universal value: Implications for international theory. In *The Environment and International Relations*, J Volger and MF Imber (eds.), 22–40. *London: Routledge.*

Homer-Dixon, T (1991). On the threshold: Environmental changes as causes of acute conflict. *International Security*, 16(2), 76–116.

Homer-Dixon, T (1994). Environmental scarcities and violent conflict: Evidence from cases. *International Security*, 19(1), 5–40.

How the Energy Sector Can Deliver on a Climate Agreement in Copenhagen; Special Early Excerpt of the World Energy Outlook 2009 for the Bangkok UNFCCC Meeting (2009). Paris: OECD/IEA.

International Energy Agency (2007). *World Energy Outlook 2007.* Paris: IEA.

Intergovernmental Panel on Climate Change, Working Group II Report (2007). Impacts, Adaptation and Vulnerability. In *IPCC Fourth Assessment Report.* http://www.ipcc.ch/

Mathews, JT (1989). Redefining security. *Foreign Affairs*, 68, 162–177.

Ministry of Economy, Trade and Industry (METI) (2006). Energy Security Study Group. *Interim Report.*

Ministry of the Environment of Japan (Sub-Committee on International Climate Change Strategy, Global Environmental Committee, Central Environment Council) (2007). *Report on Climate Security.* Ministry of the Environment.

Morgenthau, HJ (1985). *Politics Among Nations: The Struggle for Power and Peace*, 6th Ed. New York: Knopf.

Ohta, H (1998). Japan's environmental foreign policy and the concept of comprehensive national security. In *Ecological Security in Northeast*

Asia, MA Schreurs and D Pirages (eds.), 151–170. Seoul: Yonsei University Press.

Sprout, H and M Sprout (1965). *The Ecological Perspective on Human Affairs: With Special Reference to International Politics*. Princeton: Princeton University Press.

Sprout, H and M Sprout (1971). *Toward a Politics of the Planet Earth*. New York: Van Nostrand Reinhold.

Ullman, R (1983). Redefining security. *International Security*, 8(1), 162–177.

Waltz, KN (1959). *Man, the State, and War*. New York: Columbia University Press.

Waltz, KN (1979). *Theory of International Politics*. Reading, Massachusetts: Addison-Wesley.

World Commission on Environment and Development (WCED) (1987). *Our Common Future*. Oxford: Oxford University Press.

Young, OR (1992). *Arctic Politics: Conflict and Cooperation in the Circumpolar North*. Hanover and London: University Press of New England.

ENDNOTES

[1] Realists and neo-realists built their theories from roots deep in national and military security concerns. Carr, EH (1939). *The Twenty-Years' Crisis, 1919–1939: An Introduction to the Study of International Relations*. London: Macmillan.; Morgenthau, HJ (1985). *Politics among Nations: The Struggle for Power and Peace*, 6th Ed. New York: Knopf as well as Waltz, KN (1959). *Man, the State, and War*. New York: Columbia University Press; and Waltz, KN (1979). *Theory of International Politics*. Reading, Massachusetts: Addison-Wesley.

[2] Barnett, RB (1984). *Beyond War: Japan's Concept of Comprehensive National Security*. Washington, D.C.: Pergamon-Brassey's.

[3] Barnett, *ibid*, pp. 1–6.

[4] This section is a modification of my previous study (Ohta, H (1998). Japan's environmental foreign policy and the concept of comprehensive national security. In *Ecological Security in Northeast Asia*, MA Schreurs and D Pirages (eds.), pp. 151–170. Seoul: Yonsei University Press.

[5] Dyer, HC (1996). Environmental security as a universal value: Implications for international theory. In *The Environment and International Relations*, J Volger and MF Imber (eds.), pp. 22–40. London: Routledge.

[6] Homer-Dixon, T (1991). On the threshold: Environmental changes as causes of acute conflict. *International Security*, 16(2), 76–116; and

Homer-Dixon, T (1994). Environmental scarcities and violent conflict: Evidence from cases. *International Security*, 19(1), 5–40.

7 Brown, LR (1986). Redefining national security. In *State of World Environment*, Lester Brown *et al.* (eds.). New York: Norton; and Mathews, JT (1989). Redefining security. *Foreign Affairs*, 68, 162–177.

8 Ullman, R (1983). Redefining security. *International Security*, 8(1), 162–177.

9 Sprout, H and M Sprout (1971). *The Ecological Perspective on Human Affairs: With Special Reference to International Politics.* Princeton: Princeton University Press; and Sprout, H and M Sprout (1971). *Toward a Politics of the Planet Earth.* New York: Van Nostrand Reinhold.

10 Sprout and Sprout, *op. cit.* (1971, p. 14).

11 *Ibid.*, p. 25.

12 *Ibid.*, p. 18.

13 Pirages, D (1996). Environmental security and social evolution. *International Studies Notes*, 16/1, 8–13, cited by Dyer.

14 World Commission on Environment and Development (WCED) (1987). *Our Common Future*, 8. Oxford: Oxford University Press.

15 *Ibid.*

16 Deudney, D (1995). Environment and security: Muddled thinking. In *The Global Agenda: Issues and Perspectives*, CW Kegley, Jr and ER Wittkopf, 4th Ed., 446–454. New York: McGraw-Hill.

17 *Ibid.*, p. 448.

18 Young, OR (1992). *Arctic Politics: Conflict and Cooperation in the Circumpolar North*, p. 2. Hanover and London: University Press of New England.

19 WCED, *op. cit.*

20 Young, *op. cit.*, p. 215.

21 The Arctic Council was formally established by the Ottawa Declaration of 1996. This Council is "a high level intergovernmental forum to provide a means for promoting co-operation, co-ordination and interaction among the Arctic States, with the involvement of the Arctic Indigenous communities and other Arctic inhabitants on common Arctic issues, in particular issues of sustainable development and environmental protection in the Arctic", The Internet site of the Arctic Council, http://arcticcouncil.org/article/about [accessed 28 August 2011].

22 The Internet site of the Arctic Council http://arctic-council.org/section/permanent participants [accessed 28 August 2011].

23 IASC is a nongovernmental organization that aims to encourage, facilitate and promote co-operation in all aspects of Arctic research in all countries engaged in Arctic research and in all areas of the Arctic region. AMAP is one of five Working Groups of the Arctic Council and

its primary function is to advise the governments of the eight Arctic countries on matters relating to threats to the Arctic region from pollution, and associated issues. CAFF is the Biodiversity working group of the Arctic Council and its mandate is to address the conservation of Arctic biodiversity, and to communicate its findings to the governments and residents of the Arctic.

24 Regarding the following impact assessment, I referred to the Arctic Climate Impact Assessment (ACIA) (2004). *Impacts of a Warming Arctic.* Cambridge: Cambridge University Press.

25 *Ibid.*, pp. 22–33.

26 *Ibid.*, pp. 34–45.

27 *Ibid.*, pp. 58–77.

28 *Ibid.*, p. 61.

29 *Ibid.*, pp. 86–91.

30 *Ibid.*, pp. 106–111.

31 *Ibid.*, pp. 82–85.

32 Go, F, M Tukishima and Y Kato (2008). Hokkyoku Sodatsu ("The Competition for the Arctic"). the *Globe* (A special series feature articles) section, *The Asahi Shimbun*, 6 October.

33 MOFA (2009). Statement of H.E. Seiko Hashimoto, State Secretary for Foreign Affairs of Japan on the Occasion of the 50th Anniversary of the Antarctic Treaty, Washington, D.C. on 6 April. http://www.mofa.go.jp/announce/announce/2010/9/0902 01.html [accessed 25 February 2011].

34 Regarding the following descriptions about the Japanese governmental stance on the issue of the Arctic, I refer to Hidehisa Horinouchi (Deputy Director-General of MOFA's Bureau of International Legal Affairs), "Japan and the Arctic," at the Norway-Japan Polar Seminar held on 26 April 2010. The outline of his presentation is available at http://www.norway.or.jp/news events/environment/polarseminar2010/ [accessed 15 March 2011].

35 Horinouchi, *ibid.*

36 Horinouchi, *ibid.*

37 Launching of the 'Arctic Task Force (ATF). MOFA. http://www.mofa.go.jp/announce/announce/2010/9/0902 01.html [accessed 25 February 2011].

38 Chevalier, J-M (ed.) (2009). *The New Energy Crisis: Climate, Economics and Geopolitics.* New York: Palgrave Macmillan.

39 International Energy Agency (IEA) (2009). How the energy sector can deliver on a climate agreement in Copenhagen: Special early excerpt of the world energy outlook 2009 for the Bangkok UNFCCC meeting. Paris: OECD/IEA.

40 *Ibid.*

41 Japan energy data, statistics and analysis — oil, gas, electricity, coal. Energy Information Administration (EIA). file:///V:/PRJ/NewCABs/V6/Japan/Full.html [accessed 10 April 2011].

42 BP. *BP Statistical Review of World Energy June 2011* p. 41, http://www. bp. com/assets/bp internet/globalbp/globalbp uk english/reports and publications/statistical energy review 2011/STAGING/local assets/pdf/statistical review of world energy full report 2011.pdf [accessed 28 August 2011].

43 EIA, *op. cit.*

44 Ministry of Economy, Trade and Industry (METI) (2006). Energy Security Study Group. *Interim Report.*

45 *Ibid.*, p. 5.

46 *Ibid.*, pp. 2–4.

47 *Ibid.*, pp. 7–8.

48 *Ibid.*, pp. 8–9.

49 *Ibid.*, p. 9.

50 *Ibid.*, pp. 9–10.

51 *Ibid.*, p. 10.

52 *Ibid.*, pp. 11–19.

53 METI (2006). *New National Energy Strategy: Digest*, METI. http://www.enecho.meti.go.jp/english/report/newnationalenergystrategy 2006.pdf [accessed 12 April 2011].

54 *Ibid.*

55 *Nikkei Ekoloji (Nikkei Ecology)*, May 2011, p. 17.

56 The METI, "The Strategic Energy Plan of Japan: Meeting global challenges and securing energy futures (Revised in June 2010)." http://www.meti.go jp/english/press/data/20100618 08.html [accessed 28 August 2011]

57 For instance, the maximum radiation level on 6 October 2011 was 2.048 microsieverts per hour at the Iitate village office (40 km Northwest of Fukushima No. 1 Plant) while 0.067 at Shinjuku in Tokyo (Maximum radiation levels in eastern Japan (2011). *The Japan Times*, 7 October).

58 Asia Japan Watch of *The Asashi Shimbun*, "Fukushima accident raised to same level as Chernobyl." http://ajw.asahi.com/article/0311disaster/fukushima/AJ201104120020 [accessed 28 August 2011].

59 Beck, U (1992). *Risk Society: Towards New Modernity.* London: Sage.

60 EIA (2011), *op. cit.*

61 Division of Global Warming Prevention Headquarters of the Japanese government (2002). The Guideline for Measures to Prevent Global Warming, 12 March.

62 Minshuto (2009). Chikyu Ondanka-taisaku Kihon Hoan (The Democratic Party of Japan (DPJ), "The Basic Global Warming Bill"), No. 19, the 171st Diet Session, 24 April.

63 Yoshihiko Noda, Japan's Finance Minister, became the seventh prime minister in six years on 30 August 2011. His political style and idea resonate with the country's conventional prime minister of seeking a compromise within the ruling Democratic Party of Japan (DPJ) and among opposition parties (Fukue, N (2011). Noda victorious in race for prime minister: Fiscal hawk defeats Kaieda in a runoff for DPJ presidency. *The Japan Times*, 30 August).

64 One of public-opinion polls indicates that 84.8% of the respondents want to use renewable energy resources for generating electricity (wind, solar, geothermal and etc.) whereas 21.3% for nuclear energy (multiple choice answers) (JiJi Press, 20 August 2011; http://www.jiji.com/jc/zc?k=201108/2011082000191 [accessed 9 October 2011]). Similarly, another poll shows that people support the idea of reducing nuclear power plants (72%) and energy conservation despite roughing it (65%) (Mainichi seron chosa (2011). (the Mainichi public poll), 22 September. http://ratio.sakura.ne.jp/archives [accessed 9 October 2011]).

65 Buza, B, O Wæver and J de Wilde (1998). *Security: A New Framework for Analysis*, p. 23. Boulder, CO: Lynne Rienner.

66 *Ibid.*, pp. 31–32.

67 Annan, K (2006). Frightening lack of leadership on climate change in Nairobi, 15 November, SG/SM/10739, ENV/DEV/904.

68 *Ibid.*

69 Ki-moon, B (2007). Climate change requires long-term global response. The United Nations Security Council, 17 April, SG/SM/10949, SC/9001, ENV/DEV/921.

70 The Biden–Lugar Resolution, 110th Congress Session, S.RES.

71 A Transcript, 21 September 2006, Council on Foreign Relations, New York, NY. It is http://www.cfr.org/publication/1511/climate security.html [accessed 15 February 2007].

72 *Ibid.*

73 The High Representative and the European Commission to the European Council (113/08), http://www.consilium.europa.eu/ueDocs/cms Data/docs/pressData/en/reports/99387.pdf

74 Ministry of the Environment of Japan (Sub-Committee on International Climate Change Strategy, Global Environmental Committee, Central Environment Council) (2007). *Report on Climate Security*, Ministry of the Environment.

75 *Ibid.*, p. 19.

76 *Ibid.*

77 *Ibid.*, p. 21.

78 *Ibid.*, pp. 22–23.

79 Young, *op. cit.*, p. 215.

8

THE ARCTIC GOVERNANCE AND EU "SOFT POWER"

Danila Bochkarev

8.1. INTRODUCTION

The melting of the ice in the Arctic due to global warming is expected to cause major environmental and geopolitical changes in the Arctic, often also called — the High North. This phenomenon is also expected to initiate additional interests in the Arctic among the coastal states, members of the Arctic Five (A5), a group of countries that encircle the Arctic Ocean consisting of Canada, Greenland/Denmark, Norway, Russia, and the US, and also other major international powers such as China, the European Union (EU), India, and Japan. These developments create an entirely new set of polit ical, economic and social scenarios in the High North. Melting ice cover would facilitate the exploitation of energy and other mineral resources and open up access to new fish stocks. It would also open up new shipping routes, offering shorter distances for trade between Europe and East Asia. At the moment, the A5 countries are clearly the dominant economic, political and military players in the region. This informal "Arctic club" met for the first time on 27–28 May 2008 in Ilulissat, Greenland to sign the Ilulissat Declaration. This document assumed that the Law of the Sea "provides for important rights and obligations concerning the delineation of the outer limits of the continental shelf, the protection of the marine environment, including ice-covered areas, freedom of navigation, marine scientific research, and other uses of the sea."[1] We, therefore, see no need to "develop a new comprehensive international legal regime to govern the Arctic Ocean", according to the Declaration.[2] The Ilulissat Declaration also confirmed that by "virtue of their sovereignty . . . and jurisdiction in large areas of the Arctic Ocean, the five coastal states are in a unique position to address these possibilities and challenges. The Arctic Ocean is a unique ecosystem, which the five coastal states have a stewardship role in protecting."[3] The Arctic Council (AC) and the Barents Euro-Arctic Council (BEAC) were also designated by the A5 as the key institutional co-operation frameworks in the Arctic.[4] The A5 foreign ministers met again on 29 March 2010 in Chelsea, Canada to bolster regional co-operation in the Arctic. This move made by the coastal Arctic states was seen by external observers as an attempt to reassert coastal state sovereignty in the High North and exclude non-A5 countries from the Arctic regional governance. For example, Iceland's foreign affairs minister, Össur Skarphéðinsson, expressed his discontent regarding the absence of his country, as well as two other AC members — Finland and Sweden, at the Chelsea Summit.[5]

However, as the Arctic plays a significant role in global environmental stability, it is presented in many ways as a "common good", which should not be, due to its global environmental impact, made the sole responsibility of the A5.

One of the key "outsiders", the EU has never officially challenged the A5 countries' sovereign rights in the Arctic, but it has closely followed the recent political, institutional and economic developments in the High North. Brussels does realize its institutional and legal limitations. In fact, under these circumstances, it is quite difficult for the EU to become a significant "rules setting" regional actor in the Arctic. The EU, as well as its member states, does recognize the Law of the Sea and the United Nations Convention on the Law of the Sea (UNCLOS), which is considered to be the primary legal source in the Arctic. At the same time, the A5 draws its legitimacy from the same source of law.

Nevertheless, the EU, as well as other strategic players, are increasingly ambitious about raising their profile in the Arctic. The High North is one of the key issues for the EU in the area of regional governance, climate change and potential access to bio-marine resources. Since the EU is not challenging the existing legal status quo, Brussels is not likely to be directly present, at least in the near future, in setting the rules for the exploration and production of the natural resources in the Northern part of the Arctic. The bulk of energy resources and the mineral wealth in the High North are situated within the A5 respective Exclusive Economic Zones (EEZs) or their seabed territories, covered by potential extension of their continental shelf. Nevertheless, the Arctic also represents an important source of energy for the EU and Europe's dependence on gas and oil reserves is likely to grow. Even today, a significant bulk of Russian and Norwegian gas supplied to Europe comes from the hydrocarbon deposits in the High North. This dependence is likely to grow in the future.

Brussels is particularly concerned with such issues as climate change, environmental protection and natural resources management, and protection of autochthon populations. For example, a rise in global sea levels partially caused by melting of the Arctic ice cap could affect, both positively and negatively, a number of EU member states such as the Netherlands and, therefore, justifies Europe's interest in the Arctic. On the one hand, the rise of sea levels may endanger the well-being of people living along Europe's coastlines. On the other hand, the melting of the ice will have a positive effect on navigation and fisheries in the Arctic. These challenges and opportunities relate to the global aspect of the human-driven and natural regional dynamics in the High North. Therefore, they relate to the environment-driven common heritage of the humankind aspect of the Arctic.

The EU has an excellent Arctic-related research potential, which gives it a certain weight in the environment-related Arctic issues. Brussels has an extensive High North scientific program, and in the past 10 years has spent 200 million euros on Arctic research projects, which have a specific focus on the environment, pollution and rising sea levels. This scientific potential is supported by an extended policy apparatus. The EU Institutions have adopted a number of policy initiatives, setting priorities and specific action plans for EU institutions and member states in the Arctic. For example, during the first Barroso Commission (2004–2009), Energy Commissioner Andris Piebalgs and Commissioner for Maritime Affairs and Fisheries Joe

Borg both stressed the need to tap the region's natural resources. Javier Solana, former EU's High Representative for the Common Foreign and Security Policy, mentioned the Arctic in his report on climate change and international security. Last, but not least, in April 2009, the EU unsuccessfully tried to join the Arctic's most important institutional framework — the Arctic Council. The observer status is still high on the EU's agenda and remains one of the EU's top policy priorities in the High North.

The author would like to stress the peaceful, constructive and low key nature of the competition between the EU, its member states, and the Arctic coastal states. There is ongoing debate, including in Brussels, on the militarization of the Arctic. From the author's point of view, this debate is highly exaggerated and is irrelevant to the EU's policy towards the Arctic. The limited build up of the armed forces based in the Arctic is linked to the increased demand for border control facilities. The decrease in the ice cap opens new navigation routes and creates previously inexistent sea routes and borders, which need to be protected. The author believes that the theoretical competition between certain A5 countries, such as Canada and Russia, would be situated in the area of governance and sea navigation and not in the area of resource and military competition. Even debate over the use of energy resources would be held within existing legal and institutional frameworks, equally recognized by the EU. This also explains the EU's interests in extending its participation in the Arctic regional governance.

8.2. THE EU IN THE ARCTIC: HISTORICAL CONTEXT

This chapter aims to shed light on the EU's approach to the Arctic. Towards this end, understanding of the EU's Arctic political rhetoric and Brussels' motivations and interests is difficult without the historical overview of Europe's engagement with the High North, as well as the key players and major steps in the EU Arctic policy-making.

The EU's "march into the Arctic" started in 1973 when Denmark joined the European Economic Community (EEC) together with

Greenland. The majority of votes cast in Greenland went against EEC membership, nevertheless Greenland, as an inherent part of Denmark, had to join the EEC.[6] The "home rule" for Greenland was accepted in 1979 by Denmark. This document allowed Greenland to leave the then EEC in 1985, while remaining a part of Denmark.

Greenland's decision to leave the EEC left the EU with no direct access to the Arctic continental shelf. The European Community ratified the UNCLOS in April 1998 and, therefore, recognized the right of the Arctic coastal countries to full economic sovereignty in their EEZs and their right to extend their continental shelf. On a practical level, the "withdrawal decision" denied EU member states practically unrestricted access to Greenland's EEZ and its mineral and biological resources.

Greenland received autonomy under the home rule status in 1979. Further autonomy and self-rule were introduced on 21 June 2009. If massive industrial development of the natural resources deposits in Greenland becomes commercially attractive, one cannot rule out the country's eventual independence. The purely theoretical question remains open — would Greenland eventually join the EU or stay outside of the EU following Norway's example? Obviously, Greenland's EU membership is an interesting option for Brussels. It would allow Brussels, via Greenland, to join an informal club, A5 (Canada, Denmark/Greenland, Norway, Russia, and USA), and participate in the development of the Island's mineral riches.

In 1995, the EU enlargement brought three new countries to Europe: Austria, Finland and Sweden. Two of these, Finland and Sweden, are full members of the Arctic Council, the Arctic's main governing body. Therefore, the membership of Sweden and Finland created a real "Northern dimension" to European politics. The Northern Dimension policy, drafted in 1999, was one of the key priorities of the Finnish EU Presidency (July–December 2006). During the Helsinki Summit in November 2006 involving the EU, Norway, Iceland and Russia, EU leaders adopted the Northern Dimension Declaration[7] and the Northern Dimension Framework Document.[8] These agreements, detailed below, helped the EU increase its involvement in the Arctic through its Northern member states.

The Northern Dimension Framework Document is a key document which determines both the state and non-state actors' participation in this initiative and sets the key objectives. The Framework Document sets the priority co-operation sectors and provides grounds for the Northern Dimension institutional arrangements. The document also includes points on the project implementation, monitoring and financing. It also does provide provisions for eventual review of this document.[9] The Northern Dimension initiative includes the EU, Russia, Norway and Iceland and covers a broad geographical area between the Baltic Sea and the European part of the Arctic. This Initiative is aimed at providing a "common framework for the promotion of dialogue and concrete co-operation, to strengthen stability and well-being, intensify economic co-operation, and promote economic integration, competitiveness and sustainable development in Northern Europe."[10] Other important Arctic institutional players, namely, the BEAC and the AC, were invited to join the Northern partnership as full members. The Northern Dimension is, therefore, a first significant attempt made by the EU to play a more active role in the Arctic.

The implementation of the Agreement on the European Economic Area (EEA Agreement) between the EU, Iceland, Lichtenstein and Norway on 1 January 2004 was yet another important step in the EU's involvement in the High North.[11] The signature of this agreement became an important institutional victory for Brussels. The EEA Agreement transposed EU Internal Market legislation, with the exception of regulations on agriculture and fisheries, into the national legislations of the EEA Nordic members such as Norway and Iceland. Therefore, Norway, a key political and economic actor in the High North and an EEA Agreement signatory, became economically and regulatorially associated with Brussels. Oslo had to transpose the majority of EU norms and regulations, known as *acquis communautaire* into its national legislation. On 10 October 2007, the European Commission (EC) issued its EU Integrated Maritime Strategy with the reference to the Arctic made in the context of global warming.

The Strategy assumed: "attention will also be given to the geopolitical implications of climate change. In this context, the Commission will present in 2008 a report on strategic issues relating to the Arctic Ocean."[12]

On 14 March 2008, the EU High Representative and the EC issued a paper on Climate Change and International Security. This Paper directly touched upon the High North geopolitics and stressed an "increasing need to address the growing debate over territorial claims and access to new trade routes by different countries, which challenge Europe's ability to effectively secure its trade and resource interests in the region and may put pressure on its relations with key partners."[13] The EU Arctic Paper also proposed to develop "an EU Arctic policy based on the evolving geostrategy of the Arctic region, taking into account, for example, access to resources and the opening of new trade routes."[14] This document was a first attempt made by the official executive EU Institutions to reflect on Europe's role and major interests in the High North. The debate launched by the EU High Representative's Paper was continued by the European Parliament (EP) in the fall of 2008.

On 9 October 2008, the EP adopted a resolution on Arctic governance, which was seen as a first coherent attempt to start a debate on Europe's role in the Arctic's governance. The resolution urged the EC to take a proactive role in the Arctic, by becoming a permanent observer to the AC and proposed to adopt an international treaty for the protection of the Arctic, "having as its inspiration the Antarctic Treaty."[15] As a minimum starting point, such a treaty "could at least cover the unpopulated and unclaimed area at the centre of the Arctic Ocean."[16] This politically unrealistic proposal was not pursued in the following documents, but it gave additional momentum to the debate on the Arctic's governance. The EP resolution also mentioned the importance of Arctic energy resources and navigation routes for Europe and expressed the concerns about the negative environmental impacts of global warming in the High North. This resolution was a clear signal showing Europe's global ambitions and its growing involvement in Arctic affairs.

On 20 November 2008, the EC, the EU's executive body, answered the EP by publishing its own "Communication on the EU and the Arctic Region."[17] This Document confirmed, in a more diplomatic way, the EU's willingness to become an observer in the Arctic Council. The Communication proposed to "explore the possibility of establishing new, multi-sector frameworks for integrated ecosystem management."[18] In this document, the EC also touched upon EU relations with Greenland and one of the Communication's key conclusions was focused on developing closer ties with this country. In fact, Greenland's possible independence and the introduction of self rule open new opportunities for Brussels. Greenland could become a new "Arctic window" for the EU. This fact was stressed in the Commission Communication. Thus, "additional efforts should be envisaged to make the EU an even more important partner for Greenland in managing its fragile environment and the challenges confronting its population."[19]

After the financial crash in Reykjavik, Iceland, in 2008, yet another AC and EEA member has been holding talks with Brussels regarding eventual EU membership. Iceland, already strongly integrated into the EU, applied to join the EU on 16 July 2009 and negotiations formally started on 17 June 2011.

On 8 December 2009, the EU Council adopted conclusions built on the previous initiatives by the EC and the EP. The Council document made one further step in supporting Europe's interests in the High North by "recognizing Member States' legitimate interests and rights in the Arctic."[20] This document is focused on four key areas crucial for Europe's political and economic interests in the High North, i.e., freedom of navigation in the High North, environmental protection, sustainable use of natural resources and participation in the Arctic's governance.

The most recent EU document on the Arctic policy is the EP Resolution on a Sustainable EU Policy for the High North adopted on 20 January 2011. Apart from mentioning traditional Europe's concerns and interests in the Arctic, the document stressed the *common heritage of the humankind aspect* of EU Arctic Policy. The EP Resolution said that "challenges facing the Arctic are global and should therefore include all relevant actors."[21] The document also contains an important point regarding Iceland, which could also be seen as an attempt to legitimize the EU presence in the Arctic's governance. For instance, the Resolution stressed that "future accession of Iceland to the EU would transform the Union into an Arctic coastal entity… and represents a strategic opportunity for the EU to assume a more active role and contribute to multilateral governance in the Arctic region."[22] The European parliamentarians also rightfully added that "Iceland's accession to the EU would further consolidate the EU's presence in the Arctic Council."[23] The EP affirmed "its commitment not to support any arrangements, which exclude any of the Arctic EU Member States"[24] and strongly encouraged freedom of navigation in the Arctic Ocean, while asking coastal states to "refrain from introducing any unilateral arbitrary burdens, be they financial or administrative, that could hinder shipping in the Arctic, other than internationally agreed measures aimed at increasing security or protection of the environment."[25]

8.3. EU ARCTIC POLICY: MAJOR INSTITUTIONAL ACTORS

The EU Arctic policy is conducted by three major EU Institutions — the EP, the EC and the European Council. Apart from acting according to their own agendas, these European institutions also unofficially endorse the interests of certain EU member states, such as Denmark, Finland, Germany and Sweden, with key economic and political interests in the High North. The major Arctic-related policy initiatives on the EU level originate in the EP and are often driven by two political groups — the Alliance of Liberals and Democrats for Europe (ALDE)[26] and the environment-concerned Greens/European Free Alliance. Traditionally, the EP did not play a major role in elaborating the guidelines for Europe's external initiatives, leaving this role to the EC and European Council. The adoption of the Lisbon Treaty on 1 December 2009 gave the EP additional legislative power though not directly related to the area of foreign policy.[27] However, even if the EP resolutions are legally nonbinding, they have an informal political influence. The EP is, therefore, a laboratory for new policy concepts which, for a number of political reasons, cannot be directly launched by the EC or the European Council. These policy concepts are then being developed by the EC and the Council, and adopted to the current foreign policy situation.

8.3.1. *EU interests in the Arctic*

The EU's interest in the North is relatively recent and is driven by Europe's transportation and trade needs, climate change concerns, the EU's geographical position and Brussels' energy security imperative. It is influenced both by the relevant Member States (Denmark, Germany, Finland, Sweden, etc.) and EU bureaucracy itself. The High North policy is focused on four major areas:

— Protection of the fragile Arctic environment and its indigenous populations.
— Reduction of the negative effects of the global warming (the global impact of the rise in the sea levels, loss of biodiversity, migration of species, etc.) in the Arctic on European and global environmental stability.
— An attempt to use Europe's "soft power" to expand to the High North and enter formal Arctic governance frameworks, such as the AC.
— Protection of key economic interests of the member states and the EU, as a trade block (access to new sea routes, energy, mineral and bio-resources).

This analysis is, therefore, primarily focused on EU interaction with the Arctic governance frameworks and the key economic activities in the High North, representing strategic importance for EU business interests.

8.3.2. *Melting ice in the North*

The Arctic Climate Impact Assessment forecasts a rise in the Arctic's temperature of between 4°C and 7°C by the end of the 21st century, whereas the global temperature is expected to rise between 1°C and 6°C over the same timeframe.[28]

This increase in temperature could have a dramatic effect on the Arctic ice cap. Some studies are already predicting an ice-free Arctic

Ocean during the summer months by 2040. According to a recent study entitled *Snow, Water, Ice and Permafrost in the Arctic* (SWIPA) released by the Arctic Monitoring and Assessment Program (AMAP) in 2011, "the Arctic Ocean is projected to become nearly ice-free in summer within this century, likely within the next thirty to forty years."[29] The ice cap projections made by the Intergovernmental Panel on Climate Change (IPCC) in 2007 "underestimated the rates of change now observed in sea ice."[30] The report also concluded that "the largest and most permanent bodies of ice in the Arctic — multi-year sea ice, mountain glaciers, ice caps and the Greenland Ice Sheet — have all been declining faster since 2000 than they did in the previous decade."[31] A more recent study conducted by the University of Bremen in Germany stressed the "extent the Arctic sea ice has reached on 8 September 2011 with 4.240 million km^2, [being] a new historic minimum."[32] This exceeds the previous minimum of 2007, while the retreat of the summer sea ice since 1972 amounts to 50%.[33]

These estimates are often challenged by alternative studies and there is still no clear consensus on the extent to which the Arctic will be ice-free and navigable and whether the temperature rise in the Arctic is a long-term irreversible phenomenon. Nevertheless, it is obvious that even limited accessibility in the ice-free part of the Arctic represents previously unrealistic opportunities for maritime transportation and development of mineral resources, fish stocks and bio-resources.

8.3.3. *New energy province for Europe*

The US Geological Survey (USGS), *Circum-Arctic Resource Appraisal: Estimates of Undiscovered Oil and Gas North of the Arctic Circle*, released in 2008, confirmed that "90 billion barrels of oil, 1,669 trillion cubic feet of natural gas, and 44 billion barrels of natural gas liquids may remain to be found in the Arctic, of which approximately 84% is expected to occur in offshore areas."[34] Generally, the Arctic is believed to contain up to 25% of the world's remaining hydrocarbons reserves. Most of the undiscovered natural gas reserves are situated in Russia's EEZ as well as north of the Alaska coast, while oil reserves are distributed more equally. New technologies for exploring and production drilling, allowing the development of super-deep offshore hydrocarbon projects, have created new oil and gas opportunities in the High North. The recently announced partnership (August 2011) between global energy's super-major corporation, ExxonMobil, and Russia's biggest oil company, Rosneft, is a sign of the industry's growing interest in the Arctic energy deposits.[35]

The list of already discovered natural gas and oil reserves is very impressive. For example, the Shtokman gas field in the Russian part of the Barents Sea alone contains 3.9 trillion cubic meters (tcm) of gas and 53.3 million tons of gas condensate. The discovered hydrocarbon deposits in Yamal in the Northern Siberia are enormous as well. According to Gazprom's official data, Yamal oil and gas fields contain between 16 and 38 tcm of natural gas, 230.7 million tons of gas condensate and 291.8 million tons of oil.[36] According to the USGS, the Cook Inlet Region of Alaska holds "an estimated mean of 19 trillion cubic feet (or 540 billion cubic meters) of natural gas, about 600 million barrels of oil, and 46 million barrels of natural gas liquids"[37] Snøhvit offshore gas field in Northern Norway is relatively small in scale, if compared to super-giant fields in Russia. One of its relatively small reserves, discovered in 1984, contains 193 billion cubic meters of natural gas, 113 million barrels of condensate and 5.1 million tons of gas liquids.[38] However, even such a small deposit can generate 5.75 billion cubic meters (bcm) of liquefied natural gas (LNG)[39] or more than 2/3 of Singapore's annual natural gas consumption in 2010.[40]

It is still unclear, however, how fast the exploration and production of these hydrocarbon resources will take place, taking into account technological, environmental and commercial challenges. While the prospects are rather bright for small-scale offshore developments and large-onshore mega-projects, such as Yamal, the large-scale development of the underwater Arctic energy reserves will not start before 2020. The large offshore projects face serious profitability constraints during the periods of unpredictable global gas demand. Uncertain global energy supply/demand dynamics, low prices, high production costs and emergence of commercial unconventional oil and gas production have scaled down the expectations for immediate commercial development of the Arctic hydrocarbon resources. We still have to wait at least until the eventual termination of the global "gas glut"[41] in 2014–2016 in order to assess the realistic timeframe for the development of the Arctic energy projects. However, the importance of energy supplies from the High North is not to be underestimated. A significant part of Europe's post-2020 oil and gas exports will come from this part of the world. According to the *Eurogas Statistical Report* 2010, Russia and Norway accounted for, respectively, 23% and 20% of gas supplies to the EU in 2009.[42] More than 80% of Russian gas is produced in the High North and the situation is similar in Norway. In the future, the dependence of Europe on imported gas, including from the Arctic, will further increase. According to *Eurogas Long-Term Outlook 2007–2030* , the EU conventional annual natural gas production will decrease from 177 bcm in 2007 to 57 bcm in 2030. Europe will have to import between 344 and 477 bcm/year of gas, including 88 bcm from Norway.[43] A significant part of non-Norwegian natural gas will come from the Russian Arctic increasing the importance of the High North for Europe's energy security. However, the liberalization of the EU gas markets (3rd Energy Package) risks complicating Europe's relations with its Arctic suppliers, particularly Russia. On the one hand, energy producers such as Russia see the EU's attempts to break up long term contracts and energy companies' vertical integration almost as an existential threat. On the other hand, EU regulatory authorities claim that vertically-integrated energy suppliers could dominate the European markets and obstruct free competition. Needless to say, most Arctic energy projects are capital-intensive. Therefore, the investments have to be guaranteed by long-term supplies and price fluctuation hedging, which contradicts current regulatory trends in the EU. Therefore, the development of Arctic energy resources could be delayed, especially if both sides do not find a compromise.

8.3.4. *Rare earth and precious metals: a driver for the EU industry*

Apart from bio-resources and hydrocarbons, the Arctic is also rich in rare earth metals and precious metals. Rare earth metals such as lanthanum, cerium or terbium are essential for the high-tech industry and production of consumer goods such as mobile phones, computers and batteries. The global demand for rare earth metals is predicted to rise to 180,000 tons by 2012 and exceed 200,000 tons in 2014.[44]

The Arctic region is rich in rare earth metals. Greenland's largest Kvanefjeld multi-element mine is alone estimated to hold between 1.07 and 1.22 million tons of rare earth metals reserves.[45] Another example is Norilsk Nickel mines in Russia. GMK Norilsk Nikel, operating in the Russian Arctic, produces 38% of world's production of palladium and, respectively, 22% and 9% of the global nickel and platinum outputs.[46] Platinum and palladium represent a strategic interest for Europe's automobile and high-tech industries — they are used in vehicle emission control devises and electronics. The EU member states do not have their own reserves of platinum group metals. Most of these minerals are produced in Russia and South Africa. In the future, the EU rare earth/platinum metals imports from the Arctic are likely to increase.

8.3.5. *The future of EU fisheries is also in the High North*

Not only is the High North rich in mineral wealth, it is also home to important fish stocks and other bio-marine resources. Up to 25% of the global catch of whitefish is from the Arctic. The region is also abundant in invertebrates, widely used in the pharmaceutical industry. The unexploited fish reserves in the High North represent a significant interest for EU fisheries. Europe is the third largest fish producer in the world and accounts for "6% of the world's catch, worth some €6 billion a year."[47] However, the fish stocks exploited by European fleets are quickly depleting and the EU has to import significant amounts of fish, including from non-EU Arctic countries.[48] For example, fish from Norway and Iceland accounted for, 20% and 6%, respectively, of EU fish imports in 2009.[49]

In the mid-term future, climate change could provide new opportunities for fishing outside the EEZs of non-EU Arctic states. According to the United Nations Convention on the Law of the Sea (UNCLOS), coastal states do not generally own fish stocks outside their EEZs even, if their continental shelf was extended beyond the EEZ. In this particular case, the coastal states only keep control of the mineral and biological resources on or underneath the seabed, and not the bio-resources in the water column. These regulations create new opportunities for the fishing fleets from the third countries.

Most of the Arctic is not covered by existing international fisheries rules and the absence of fishing regulations in the Northern part of the Arctic high seas waters is a matter of concern for Brussels as this might "lead to unregulated fisheries."[50] The EU official statement stresses the fact that "the European Community is a member of the North East Atlantic Fisheries Commission (NEAFC). It cooperates with the states "having sovereignty or jurisdiction in Arctic waters, seeking not only to ensure fishing opportunities, but also to guarantee long-term conservation and optimum utilization of fishery resources."[51] Apart from the environmental sustainability aspect, the EU is unofficially concerned about future attempts by one of the coastal Arctic states, or a combination of the A5 countries, to introduce a new fisheries regime in the seas not covered by the current regulations and deny them access to the fish stocks. Fishing in the High North, therefore, risks becoming a latent, yet very important, food security issue.

8.3.6. *A new maritime highway would be crucial for EU–Asia trade*

Last, but not the least, the High North is becoming a significant global maritime highway. The Arctic has two major sea routes — the Northern Sea Route/"Sevmorput" (NSR) in the Russian Arctic, and Canada's North-West Passage (NWP). The NSR route, for instance, cuts the distance between Rotterdam and Yokohama by 35% from 13,000 to 8,500 nautical miles (or from 24,000 to 15,750 km) and theoretically allows ship operators to save time and fuel. For example, Russian independent gas producer Novatek has been shipping gas condensate via the NSR since August 2010. Leonid Mikhelson, CEO of Novatek, stated that in 2011 his company would be able to "save at least 10–15% on transportation costs [on cargo deliveries] to Pacific-Asian markets."[52]

The issue of maritime navigation and container traffic is of crucial importance for the EU, the world's leading sea power. According to EC statistics, 90% of external and 43% of domestic EU trade is transported by sea and around 1 billion tons of freight are "unloaded and loaded annually in Union ports."[53] Shipping companies owned by EU businesses "control one third of the world's fleet, and around 40% of EU trade is carried on vessels controlled by EU interests."[54] German shipping operators alone control the largest container fleet worldwide, comprising over 30% of global container capacity.[55]

Therefore, the NSR and the NWP could potentially play a significant role in EU-Asia trade. The cargo flows between Europe and Asia are rapidly increasing, especially the container cargo shipments segment. In July 2011, the Suez Canal traffic reached 78.83 million tons with container ships traffic accounting for 46.457 million tons.[56] An additional increase in the container traffic would create serious congestion problems in the Suez Canal. In this context, the NSR and the NWP could offer good mid-term shipping opportunities, especially for the medium-scale intermodal container navigation.

Currently, the NSR route is ice-covered most of the year and the cargo ships almost always have to be guided by icebreakers, which adds significantly to the shipping costs. A similar situation is expected for the NWP. In 2011, the NSR was open for navigation between June and October. However, in 20–30 years time, with the reduction of the Arctic's ice cap, a typical passage during the summer navigation period would not need an icebreaker escort.

Canada's NWP, passing North of Canada's mainland and connecting the Atlantic and Pacific Ocean, is yet another shipping route. For example, usage of the NWP cuts the London-Tokyo route from 23,000 to 16,000 km. The NWP is not widely used for maritime navigation as this passage lacks necessary ground infrastructure and icebreaker capacity. However, in August 2007, Canadian Prime Minister Stephen Harper announced his country's plans to build its first Arctic deep-sea port to bolster Canada's sovereignty and infrastructure in the Arctic.[57]

The EU is trying to reassert its position as a global maritime actor and maintain the principles of free navigation. For instance, while Ottawa considers the NWP a part of Canadian internal waters, the US and the EU claim the NWP should be treated, according to the UN Convention on the Law of the Sea (UNCLOS) as a transit passage, allowing free and unrestricted traffic. The importance of maritime transportation for the EU trade explains Brussels position on the NWP and Europe's interest in the NSR. Both Arctic sea paths represent a new transport solution for Euro-Asia cargo and could, in the mid-term future, contribute to EU sea trade with Asia. Therefore, it is in the EU's interests to find common ground with the transit countries — Canada and Russia.

In order to fully support its regional interests, the EU should participate in the Arctic's regional governance. Until recently, Brussels' presence in the High North's institutional frameworks was rather limited. Nevertheless, the EU has never stopped trying to join, at least as a permanent observer, the Arctic's most promising organization — the Arctic Council. Therefore, the Law of the Sea and the regional co-operation frameworks are crucial for understanding the EU positioning in the Arctic regional governance.

8.3.7. *The European Union, Arctic and the Law of the Sea*

The UNCLOS, also known as the Law of the Sea, is the main source of international law for the Arctic. Concluded in 1982, it came into force in 1994 and was ratified by all A5 states, except the US. The EU, as a legal entity, and all EU member states, are part of the Convention as well. The Law of the Sea provides comprehensive guidance on a large number of issues connected to the use of ocean space and its resources. For instance, the UNCLOS defines the EEZ and continental shelf of a coastal state and provides guidance on navigation and extraction of mineral and biological resources. The usage of natural reserves in the EEZs, navigation through the straits in the High North, and extension of a continental shelf are possibly the three most relevant issues relating to the application of the Law of the Sea in the Arctic Ocean.

The UNCLOS gives the coastal states the right to a 200-mile EEZ, within which the state maintains an exclusive mandate to develop mineral and biological resources in the waters and in the subsoil of the continental shelf. The seabed outside national EEZs is controlled by the UN International Seabed Authority (ISA). The ISA has the power to sanction exploration and mining. The Law of the Sea also gives additional sovereignty to the coastal states by enabling them to use the UNCLOS environment protection provisions. For example, Article 234 of the Convention allows signatories to enact special legislation to protect ice-covered waters within their EEZ and, therefore, limit potentially harmful economic activities. The continental shelf can be extended beyond the national EEZ if a claimant state can prove to the UN Commission on the Limits of the Continental Shelf (UNCLCS) that the geological structure of the continental shelf is an extension of its continental platform.

The UNCLOS also governs transit passage regimes in the context of traditional legal concepts such as "freedom of navigation" and "innocent passage through territorial waters."[58] The international straits are defined as "straits which are used for international navigation between one part of the high seas or an EEZ and another part of the high seas or an exclusive economic zone."[59] The regime of transit passage retains the international status of straits and gives the right to unimpeded navigation. Vessels in transit passages must, however, observe international regulations on navigational safety, civilian air-traffic control, and prohibition of vessel-source pollution.[60]

For example, Canada's NWP is de facto described in the UNCLOS Article 38 maintaining that

> *"In straits referred to in Article 37, all ships and aircraft enjoy the right of transit passage, which shall not be impeded; except that, if the strait is formed by an island of a State bordering the strait and its mainland, transit passage shall not apply if there exists seaward of the island a route through the high seas or through an EEZ of similar convenience with respect to navigational and hydrographical characteristics."*[61]

Based on different interpretations of the UNCLOS provisions, the EU and the US claim that the NWP regime should be governed according to the UNCLOS provisions on the transit passage, while Ottawa says it must be treated as Canada's internal waters.[62]

The application of the UNCLOS provisions in the High North could lead to geopolitical changes with a potentially adverse effect for the non-Arctic states' resource interests in the High North. The demands of A5 countries to extend their continental shelf, if accepted by the UNCLCS, could leave only a tiny portion of the Arctic under ISA governance. If so, virtually all prospective mineral and biological resources on and underneath the Arctic seabed would belong exclusively to the A5 countries. These provisions explain the EU's interest in following the activities of the UNCLCS.

8.3.8. *The EU and Arctic institutional governance*

The AC and the BEAC are the two most important components of the Arctic institutional framework. The EU is a founding member of the BEAC, but is yet to get an observer status in the Arctic Council.

The AC is the only major intergovernmental framework in the High North involving all the A5 states. Therefore, it unofficially serves as a main discussion forum on High North issues. The Council was established following the Ottawa Declaration of 1996 as a high-level inter-governmental forum addressing the issues of environmental protection and sustainable development of the polar region.[63] The AC is a forum without legal personality, and thus not a proper international organization under international law.

The AC involves all A5 states as well as certain EU member states, Finland, Iceland and Sweden. Organizations, representing indigenous populations, have been recognized as "Permanent Participants". They participate in all aspects of the AC work, but do not vote. Six non-Arctic countries (France, Germany, Netherlands, Poland, Spain, and the United Kingdom) have been admitted as Permanent Observer States to the Arctic Council.[64] At the end of 2010, nine international organizations, such as the United Nations Development Program and United Nations Environment Program, were Council observers as well.[65]

The AC is governed via the Council Chairmanship, which rotates every two years, and ministerial meetings, held by the countries holding the AC chairmanships.[66] The Council consists of six Working Groups formed by representatives of national governments of the AC member states with participation of representatives of the Permanent Participants. All decisions of the AC are made by consensus of its eight full members. The Secretariat, hosted by the Norwegian Polar Institute, is situated in Tromsø. In 2009, the EU unsuccessfully attempted to join the AC as a permanent observer. Indeed, getting a permanent observer position is still one of the EU's key "Arctic" priorities.

During the ministerial meeting in Nuuk, Greenland, held 11–12 May 2011, AC foreign ministers adopted the first legally binding agreement negotiated under the auspices of the AC — *Search and Rescue in the Arctic* (SAO).[67] The SAO could influence key business interests of the EU member states such as navigation and civil aviation flights over the North Pole. Without any doubt, this is a first step for the AC to become a real policy forum in the High North. This trend explains why the EU, eager to play a bigger role in global affairs, is trying to get access to the Arctic Council. Even a permanent observer status in the AC would help Brussels participate in the development of regional norms and regulations.

Though Brussels does not directly participate in AC activities, it is involved in the work of another Arctic organization — the BEAC. The BEAC was established in 1993 by Iceland, Denmark, Norway, Sweden, Finland, Russia, and the EC to foster regional co-operation and reduce environmental threats and security challenges.[68] The primary goal of the BEAC is to promote sustainable economic and social development in the Barents Region and thus contribute to peaceful development in the northernmost part of Europe. It initiates cross-border co-operation, which benefits from EU regional assistance, and co-ordinates environmental programs together with the EU in the framework of its Northern Dimension initiative.

Cooperation in the Barents has been institutionalized on two levels: the intergovernmental level of the BEAC and the interregional level as embodied by the Barents Regional Council (BRC). Where appropriate, there is also co-ordination with the relevant activities of the Nordic Council of Ministers, the Council of Baltic Sea States and the AC. The Barents co-operation has been a successful co-operation tool on a wide range of fields, including the indigenous peoples and a multitude of specialist, business and people-to-people networks.

Similar to the AC, the BEAC does not have binding decision-making powers and is therefore, to a large extent, unable to settle disputes or enhance co-operation. In addition, it includes only three of the five Arctic surrounding states as full members; Canada and the US hold only an observer status. The BEAC's limited geographical scope and responsibilities significantly reduce its strategic influence in the High North.

The recent emergence of the AC with its first legally binding initiative risk further reducing BEAC's role in the Arctic and the significance of this organization for the EU Arctic policy. It is clear that in the near future, the AC will become the only significant co-operation framework in the High North. It, therefore, deflates EU membership in the BEAC and increases the relevance of the AC for the EU.

8.4. CONCLUSION

The role of the EU as an "agenda-setter" in the Arctic will remain limited in the near future unless Greenland comes back to the EU. An eventual EU membership for Greenland will open an "Arctic door" for Brussels. Nevertheless, Brussels is getting ready for the ice-free Arctic in the future and is trying to get involved in regional governance despite the hidden opposition from certain A5 members such as Canada and Russia. The EU will, therefore, use both formal and informal tools to make its voice heard in the Arctic.

Europe's interest in the High North is conditioned by the growing economic and environmental importance of the Arctic region. Global environmental degradation, especially the negative impact of greenhouse gas emissions, is an issue of major concern for the EU, as well as for its member states. The EU took legally binding commitments with its 20% reduction target for emissions by 2020. The increase in public support for the "green" agenda among EU citizens strongly influences the political agenda in Brussels, particularly in the EP. It forces the parliamentarians and officials to present the EU as a global environmental champion. This fact partly explains Brussels' stance on the Arctic issues.

Direct challenges to Europe's environmental security are yet another driver for EU Arctic policy. The Arctic's melting ice cap and resulting rise in sea levels will have serious global environmental, economic, and human security ramifications outside the Arctic region. These negative trends could affect the well-being of EU citizens living in the coastal countries, such as the Netherlands, and, therefore, require a proper policy response.

The Arctic also brings new economic opportunities not to be missed. It opens new sea routes to Asia and offers a vast choice of mineral wealth, energy and bio-resources. In the mid-term, use of these resources will be necessary in order for EU businesses to conserve their global competitive edge.

This complex set of challenges and opportunities will continue defining the EU's Arctic policy in the mid-term. In this particular case, the EP will likely serve as an "incubator" and testing ground for new policy ideas. On a more practical level, (new) policy mechanisms and proposals will likely be jointly developed and implemented by the European Council, EC and interested EU member states. The EU member states do not lay any formal territorial claims in the Arctic and none of them is proceeding with such claims. They recognize the A5 countries' sovereignty over their natural resources in their respective EEZs. However, the EU institutions and member states are more assertive in promoting unrestricted navigation in this area.

They are also eager to extend their participation in the regional governance frameworks. The potential tensions, in this case could be in a purely institutional domain: the A5 would try to limit the accession of the EU, as well as other non-Arctic countries, to the regional governance mechanisms.

BIBLIOGRAPHY

A Maritime Policy for the EU (2007). 10 October. http://ec.europa.eu/maritimeaffairs/subpage en.html

BP (2011). *BP Statistical Review of World Energy.* http://www.bp.com/sectionbodycopy.do?categoryId=7500&contentId=7068481

Charron, A (2005). The Northwest passage in context. *The Canadian Military Journal*, 6(4).

Climate Change and International Security (2008). Paper from the High Representative and the European Commission to the European Council, 14 March. www.consilium.europa.eu/uedocs/cms data/docs/pressdata/.../99387.pdf

Communication from the Commission to the European Parliament and the Council (2008). *The European Union and the Arctic Region*, 20 November. http://eeas.europa.eu/arctic region/docs/com 08 763 en.pdf

Council conclusions on Arctic issues (2009). 8 December. http://ec.europa.eu/maritimeaffairs/pdf/arctic council conclusions 09 en.pdf

EU Maritime Policy actions: Fisheries. http://ec.europa.eu/maritimeaffairs/arctic fisheries en.html [accessed on 5 September 2011].

Eurogas. *Eurogas Statistical Report* 2010. http://www.eurogas.org/uploaded/Eurogas%20Statistical%20Report%202010Final%20291110.pdf

Eurogas. *Eurogas Long Term Outlook for Gas Demand and Supply 2007–2030.* http://www.eurogas.org/uploaded/Eurogas%20LT%20Outlook%202007-2030 Final 251110.pdf

European Parliament resolution of 9 October 2008 on Arctic governance. http://www.europarl.europa.eu/sides/getDoc.do?type=TA&reference=P6-TA-2008-0474&language=EN

European Parliament resolution of 20 January 2011 on a sustainable EU policy for the High North. http://www.europarl.europa.eu/sides/getDoc.do?pubRef=-//EP//TEXT+TA+P7-TA-2011-0024+0+DOC+XML+V0//EN

Greenland Minerals and Energy Ltd. Company data on the Kvanefjold project (June 2009 estimates). http://www.ggg.gl/Projects/Kvanefjeld-Project-Greenland.htm

Harper Announces Northern Deep-Sea Port, Training Site (2007). CBC News, 11 August. http://www.cbc.ca/news/canada/story/2007/08/10/port-north.html

Hassol, S (2004). *Impacts of Warming Arctic, Arctic Climate Impact Assessment, Executive Summary.* Cambridge: Cambridge University Press.

Heygster, G (2011). Arctic Sea Ice Extent Small as Never Before. Press Release by the University of Bremen, Germany, 8 September. http://www.iup.uni-bremen.de:8084/amsr/minimum2011-en.pdf

Humphries, M (2010). *Rare Earth Elements: The Global Supply Chain.* Congressional Research Service.

Iceland upset by Arctic summit snub (2010). *CBCNews World*, 16 February. http://www.cbc.ca/news/world/story/2010/02/16/north-arctic-five.html

Nash, ML (1997). The European Union and Secession. *Contemporary Review*, 270(1576), 247–251.

Norilsk Nikel company data. http://www.nornik.ru/investor/fact/

Northern Dimension Policy. *European External Action Service.* http://eeas.europa.eu/north dim/index en.htm

Political Declaration on the Northern Dimension Policy made by the European Union, the Republic of Iceland, the Kingdom of Norway, and the Russian Federation on 30 January. http://eeas.europa.eu/north dim/docs/pol dec 1106 en.pdf

Rudloff, B (2010). *The EU as a Fishing Actor in the Arctic: Stocktaking of Institutional Involvement and Existing Conflicts.* Working Paper, FG2, SWP Berlin.

Russia's Novatek to Ship 6–8 Cargoes via Arctic to Asia in 2011 (2010). *Platts*, 26 August. http://www.platts.com/RSSFeedDetailedNews/RSSFeed/HeadlineNews/NaturalGas/8909460

Significant Natural Gas Resources Remain to Be Discovered in Cook Inlet, Alaska (2011). *USGS Press Release*, 28 June. http://www.usgs.gov/newsroom/article.asp?ID=2841

Snow, Water, Ice and Permafrost in the *Arctic, SWIPA* — 2011. http://amap.no/swipa/SWIPA2011ExecutiveSummaryV2.pdf

Suez Canal Traffic Statistics, July 2011. http://www.suezcanal.gov.eg/TRstat.aspx?reportId=1

The Arctic Council. www.arctic-council.org/

The Barents Euro-Arctic Council. www.beac.st/

The European Commission. http://ec.europa.eu/index en.htm

The European Council. www.consilium.europa.eu/

The European Parliament. www.europarl.europa.eu/

The Ilulissat Declaration adopted by A5 Foreign Ministers (Undersecretary of State for the United States) on May 28 at Arctic Ocean Conference, Ilulissat Greenland, 27–29 May 2008. http://www.oceanlaw.org/downloads/arctic/Ilulissat Declaration.pdf

The Treaty of Lisbon. http://europa.eu/lisbon treaty/index en.htm

The United Nations. www.un.org

The United States Geological Survey (USGS). www.usgs.gov

United Nations Convention on the Law of the Sea of 10 December 1982. http://www.un.org/depts/los/convention agreements/texts/unclos/UNCLOS-TOC.htm

ENDNOTES

[1] The Ilulissat Declaration adopted by A5 Foreign Ministers (Undersecretary of State for the United States) on May 28 at Arctic Ocean Conference, Ilulissat Greenland, 27–29 May 2008. http://www.oceanlaw.org/downloads/arctic/Ilulissat Declaration.pdf [accessed 8 September 2011].

2 *Ibid.*

3 *Ibid.*

4 The Arctic governance and regional organizations are discussed in the chapter on the EU and Arctic Institutional Governance.

5 Iceland Upset by Arctic Summit Snub (2010). *CBCNews World*, 16 February. http://www.cbc.ca/news/world/story/2010/02/16/north-arctic-five.html [accessed 8 September 2011].

6 Nash, ML (1997). The European Union and Secession. *Contemporary Review*, 270.

7 Political Declaration on the Northern Dimension Policy made by the European Union, the Republic of Iceland, the Kingdom of Norway, and the Russian Federation on 30 January. http://eeas.europa.eu/north dim/docs/pol dec 1106 en.pdf [accessed on 1 August 2011].

8 Northern Dimension Policy. *European External Action Service*. http://eeas.europa.eu/north dim/docs/frame pol 1106 en.pdf [accessed on 1 August 2011].

9 For more details, please see full text of the Northern Dimension Policy Framework. http://eeas.europa.eu/north dim/docs/frame pol 1106 en.pdf

10 Northern Dimension Policy. *European External Action Service*, http://eeas.europa.eu/north dim/index en.htm [accessed on 9 August 2011].

11 For more information, please visit the EEA section of the European Union External Action Service. http://eeas.europa.eu/eea/ [accessed on 12 September 2011].

12 An Integrated Maritime Policy for the European Union, adopted on 10 October 2007 by the European Commission. http://www.dnv.com/moreondnv/research innovation/foresight/outlook/index.asp [accessed on 7 September 2011].

13 Climate Change and International Security (2008). Paper from the High Representative and the European Commission to the European Council, 14 March. www.consilium.europa.eu/uedocs/cms data/docs/pressdata/.../99387.pdf [accessed on 11 August 2011].

14 *Ibid.*

15 European Parliament resolution of 9 October 2008 on Arctic governance. http://www.europarl.europa.eu/sides/getDoc.do?type=TA&reference=P6-TA-2008-0474&language=EN [accessed on 14 August 2011].

16 *Ibid.*

17 Communication from the Commission to the European Parliament and the Council (2008). The European Union and the Arctic Region, 20 November. http://eeas.europa.eu/arctic region/docs/com 08 763 en.pdf [accessed on 30 August 2011].

18 *Ibid.*

19 *Ibid.*

20 Council conclusions on Arctic issues, 8 December 2009. http://ec.europa.eu/maritimeaffairs/pdf/arctic council conclusions 09 en.pdf [accessed on 1 September 2011].

21 European Parliament resolution of 20 January 2011 on a sustainable EU policy for the High North, Point 46. http://www.europarl.europa.eu/sides/getDoc.do?pubRef=-//EP//TEXT+TA+P7-TA-2011-0024+0+DOC+XML+V0//EN [accessed on 10 September 2011].

22 *Ibid.*, point 6.

23 *Ibid.*

24 *Ibid.*, point 45.

25 *Ibid.*, point 14.

26 For example, the ALDE initiated 2008 EP Resolution on the Arctic.

27 For more details on the EU's fundamental treaty see http://europa.eu/lisbon treaty/index en.htm

28 ACIA (2004). *Impacts of Warming Arctic, Arctic Climate Impact Assessment, Executive Summary*, p. 16. Cambridge: Cambridge University Press.

29 Snow, Water, Ice and Permafrost in the Arctic (SWIPA-2011/Executive Summary). http://amap.no/swipa/SWIPA2011ExecutiveSummaryV2.pdf [accessed on 5 September 2011], p. 7.

30 *Ibid.*, p. 7.

31 *Ibid.*, p. 6.

32 Heygster, G (2011). Arctic Sea Ice Extent Small as Never Before. Press Release by the University of Bremen, Germany, 8 September. http://www.iup.uni-bremen.de:8084/amsr/minimum2011-en.pdf [accessed on 10 September 2011].

33 *Ibid.*

34 USGS, *USGS Fact Sheet 2008–3029.* http://pubs.usgs.gov/fs/2008/3049/fs2008-3049.pdf [accessed on 5 September 2011].

35 Rosneft and ExxonMobil to Join Forces in the Artic and Black Sea Offshore, Enhance Co-operation Through Technology Sharing and Joint International Projects (2011). Rosneft Press Release, 30 August. http://www.rosneft.com/news/pressrelease/30082011.html [accessed on 7 September 2011].

36 Gazprom's Yamal mega-project description. http://www.gazprom.com/production/projects/mega-yamal/ [accessed on 7 September 2011].

37 Significant Natural Gas Resources Remain to Be Discovered in Cook Inlet, Alaska (2011). *USGS Press Release*, 28 June. http://www.usgs.gov/newsroom/article.asp?ID=2841 [accessed on 2 September 2011].

[38] For more details, please see http://www.offshore-technology.com/projects/snohvit-field/ [accessed on 7 September 2011].

[39] *Ibid.*

[40] For more details see *BP Statistical Review of World Energy*, June 2011, p. 23.

[41] An expression used to indicate oversupply of natural gas in the global gas markets.

[42] Eurogas Statistical Report 2010. http://www.eurogas.org/uploaded/Eurogas%20Statistical%20Report%202010Final%20291110.pdf [accessed on 12 September 2011].

[43] Long Term Outlook For Gas Demand and Supply 2007–2030 http://www.eurogas.org/uploaded/Eurogas%20LT%20Outlook%202007-2030 Final 251110.pdf [accessed on 12 September 2011].

[44] Humphries, M (2010). *Rare Earth Elements: The Global Supply Chain.* Congressional Research Service, 28 July.

[45] Data provided by Greenland Minerals and Energy Ltd., the operator of Kvanefjeld project (June 2009 estimates). http://www.ggg.gl/Projects/Kvanefjeld-Project-Greenland.htm [accessed on 5 September 2011].

[46] The Company data available at http://www.nornik.ru/investor/fact/ [accessed on 5 September 2011].

[47] Facts and figures on the Common Fisheries Policy. Basic statistic data. European Commission, DG Maritime Affairs and Fisheries, 2010. http://ec.europa.eu/fisheries/documentation/publications/pcp en.pdf [accessed on 7 September 2011].

[48] *Ibid*

[49] Statistics Compiled by Rudloff, B (2010). *The EU as a Fishing Actor in the Arctic: Stocktaking of Institutional Involvement and Existing Conflicts*, Working Paper, FG2, SWP Berlin, 12.

[50] EU Maritime Policy actions: Fisheries. http://ec.europa.eu/ maritimeaffairs/arctic fisheries en. html [accessed on 5 September 2011].

[51] *Ibid.*

[52] Quoted in Russia's Novatek to ship 6–8 cargoes via Arctic to Asia in 2011 (2010). *Platts*, 26 August. http://www.platts.com/RSSFeed-DetailedNews/RSSFeed/HeadlineNews/NaturalGas/8909460 [accessed on 28 August 2011].

[53] For more details, see ec.europa.eu/maritimeaffairs/press/pdf/fiche3 en.pdf [accessed on 28 August 2011].

[54] *Ibid.*

[55] Data available on http://ec.europa.eu/maritimeaffairs/pdf/country factsheets/germany en.pdf [accessed on 28 August 2011].

[56] *Suez Canal Traffic Statistics*, July 2011. http://www.suezcanal.gov.eg/TRstat.aspx?reportId=1 [accessed on 8 September 2011].

[57] Harper announces Northern deep-sea port, training site (2007). *CBC News*, 11 August. http://www.cbc.ca/news/canada/story/2007/08/10/port-north.html [accessed on 2 October 2011].

[58] *The United Nation Convention on the Law of the Sea*, Articles 34–45.

[59] *Ibid.*, Article 37.

[60] *Ibid.*, Article 19.

[61] *Ibid.*, Article 38.

[62] For more details, please see Charron, A (2005). The Northwest passage in context. *The Canadian Military Journal*, 6(4), 3.

[63] For more details, please see the Arctic Council. http://arctic-council.org/article/about

[64] *Ibid.*

[65] For more details on the Arctic Council observers, please see http://arctic-council.org/section/observers

[66] For more information on the Arctic Council, please see the Organization website. http://www.arctic-council.org/

[67] For more details, please see the agenda of Nuuk Ministerial Meeting of the Arctic Council. http://arctic-council.org/filearchive/Arctic%20Council%20Ministerial%20Meeting%20-%20DRAFT%20AGENDA.pdf

[68] For more details, please see the official homepage of the Barents Euro-Arctic Council. http://www.beac.st/in English/Barents Euro-Arctic Council.iw3

9

CONCLUSION

Hooman Peimani

Energy is a necessity for the normal operation, economic growth, prosperity and security of all countries, large and small, alike. This reality has become especially apparent over the last two centuries, which have been characterized by phenomenal industrial and agrarian growth and extension of services to which the enlargement and sophistication of security forces must be added. All these material achievements have been energy-intensive, leading to an astronomical global demand for energy (12,002.4 million tons of oil equivalent in 2010)[1] compared to the rest of human history. Given the crucial role of energy in all countries, particularly those having large and growing economies and/or populations, the quest for energy security has been a major preoccupation of their governments. Indeed, energy has become a key factor in shaping many countries' foreign and defence policies. However, the material gains of the last two centuries have been achieved at a very high price. Human endeavours in this regard have resulted in enormous emissions of greenhouse gases (GHG) in which the share of CO_2 is the largest. The latter's main source has been fossil fuels, which have resulted in degradation of the environment in different ways. Global warming is the single most destructive consequence of mankind's gluttonous energy consumption habits over the past two centuries, whose continuation will challenge the survival of all forms of life on Planet Earth.

Global warming has taken several forms of which the melting of the Arctic ice is the most important. Started a few decades ago, the gradual process has shrunk a significant part of the frozen Arctic Ocean,though the worse is yet to come. Functioning as a major cooler for the Earth, the Arctic's gradual disappearance due to melting, if it continues, will surely have grave consequences for the blue planet, including its further warming and affecting all aspects of life in one form or another and causing rising sea levels. However, in the interim, the warming of the Arctic makes at least parts of its mineral and energy resources accessible. Moreover, the same environmental trend will lead to the formation of two sea routes via the Arctic navigable for a few months initially and throughout the year in the case of a complete melting of the Arctic Ocean. Being much shorter than the route via the currently used Suez Canal and the Straits of Malacca, the Northern Sea Route (NSR) and the Northwest Passage (NWP) will significantly reduce the cost of intercontinental shipping and thus make products exported through them cheaper compared to the same products carried via the Southern Sea Route.

Clearly, the economic prospects, though arising from a catastrophic environmental trend, have raised the Arctic countries' stakes in accessing the region's resources and securing navigation channels. Of course, interest in the region is not confined to the Arctic states, as large non-regional trading countries could also benefit from the Arctic resources, particularly, its oil and gas reserves. Once developed, these reserves can help them meet a portion of their huge and growing energy requirements and thus allay their energy security concerns. Needless to say, they could also benefit from the NSR and the NWP as trading nations. The mentioned economic gains from the melting ice could also unleash tensions, disputes and conflicts between and among all of these stakeholders and justify a degree of militarization, a process already in place, the extent of which varies from one country to another as determined by each country's overall economic and military capabilities and the relative importance of its stake.

To put these issues in context, all the aforementioned potential challenges and opportunities hinge on the continuation of the current melting trend caused by global warming. It will determine whether or not these challenges and opportunities could become realities. Shawn Marshall and Vyacheslav Mikhailovich Makeev offer two different views, which represent the two major assessments in the ongoing climatic change debate. In fact, they reflect the two opposite views of the broader issue of global significance, namely whether or not global warming is actually taking place, and if so, whether it is human-made or natural. Basing his assessment on empirical evidence, Shawn Marshall attributes the ongoing melting of the Arctic ice to global warming caused by human-made emissions of GHG. Thus, the melting process will continue unless global warming is stopped. Otherwise, dire environmental consequences will be inevitable, including worsening global warming and rising sea levels. Vyacheslav Mikhailovich Makeev's view is based on a different argument based on global patterns of climate change, which have been in place for a long time, with alternating cooling and warming periods. According to him, the main influences are terrestrial and celestial. He does not see any pattern of melting affecting the entire Arctic and contends that while parts of it are experiencing melting, others are experiencing cooling.

For the good of our planet, one hopes that the total or extensive melting of the Arctic ice will not happen in the foreseeable future. It is an environmentally catastrophic prerequisite for accessing and developing the bulk of the region's mineral and fossil energy resources (mainly oil/gas) and making available the ice-free navigation of the NSR and NWP for a few months each year or the entire year. In the meantime, the development of a part of the regional resources will be feasible. In fact, some of the regional energy resources are currently being developed (e.g., gas fields in the Russian Arctic sector). Navigation for more than a few weeks a year and for all types of cargo ships and tankers is not yet possible, though predictions have been made that it might be possible as early as 2013. Nevertheless, within the framework of the current reality, the Arctic is gaining importance to its littoral states, of course to varying extents, not only because of its mineral and energy resources and shipping potential, but also its military/security significance. Section II reflects this reality by concentrating on two countries, Russia and Norway.

Nadezhda Klimovna Kharlampyeva and Arild Moe argue that the region does not necessarily have to turn into a scene of conflict even though it may be prone to it.

In her analysis of the factors, which make the region important for Russia, Nadezhda Klimovna Kharlampyeva traces Russian interests in the Arctic back to the Soviet era. Undoubtedly, the energy and mineral resources of the Russian sector are a major and, perhaps, the single major factor creating a strong stake for Russia in the Arctic. In fact, Russia's major oil and gas fields are located in and around its Arctic sector and most of the country's large exports of oil and gas are from there. These exports account for the bulk of Russia's export revenues and will remain so in the predictable future. Yet, Russia's Arctic energy resources require heavy investments in order to be fully exploited, and it is yet unclear whether it will be Russian or foreign corporations which invest the most. With a major proportion of the Arctic region's minerals, the Russian Arctic sector is important for Moscow, equally as a supplier of certain minerals for the Russian economy and as source of foreign export-generated income potentially rivalling the revenues derived from the export of energy. Despite the undeniable importance of the region's mineral and oil and gas reserves, as well as the potential Northern Sea Route, Kharlampyeva convincingly argues the importance of the Arctic in terms of military/security factors. Housing some of the Russian naval and aerial assets, the Arctic is considered a line of defence for Russia while it provides the Russian navy with access to the open seas. Hence, Russia is strengthening its military presence in the Arctic. Having said this, and acknowledging the potential for conflict between Russia and other Arctic littoral states, Kharlampyeva opines that co-operation between these states would be a possibility, thanks to a variety of factors, including their membership in certain regional organizations such as the Arctic Council and their shared concerns about the necessity to preserve the Arctic's environment. Hence, it is possible that conflicts over the region could be avoided.

Arild Moe views the Arctic as being equally important to Norway, for its economic, energy, trade and military/security dimensions as well as its undeniable environmental importance. The environmental degradation resulting from the melting of the ice will certainly have immediate negative implications for Norway as a littoral state, as it will have for the other Arctic states. The Norwegians are especially cautious about extensive engagement in the Arctic, mindful of its potentially worsening implications for the region's fragile environment. This factor, according to Moe, has been a major barrier to the development of the oil and gas resources in the Norwegian Arctic sector, and will slow down its exploitation of such resources in the near future. This is contrary to the Russian case where the environmentalists do not have a strong say in planning the development of their countries' Arctic sector. Although the environmental factor has been a barrier to accessing the regional resources, Moe stresses the crucial importance of Arctic oil and gas reserves for Norway because its current role as a major global oil and gas exporter will end within the next two decades due to the rapid depletion of its North Sea oil and gas reserves. Hence, developing the Norwegian oil and gas reserves in its Arctic sector is essential for maintaining its exporting status. Among others, this factor is a major incentive for Norway to ensure peace and stability in the Arctic as reflected in the peaceful settlement to date of almost all its disputes with Russia over division of the Arctic's Barents Sea, leaving only one zone to be divided between them. While the Arctic is important for all its states, especially Russia, Moe does not see conflict as an inevitable scenario given that the harsh environmental reality of the Arctic sets limits to regional ambitions, and that a degree of regional co-operation has already been in place for a while. This is also true with respect to Norwegian–Russian ties and even their co-operation in developing some Russian energy fields. In short, he does not anticipate major conflicts arising from disputes over the regional resources, at least in the foreseeable future, though the seeds of conflicts do exit.

Without doubt, the main beneficiaries of the Arctic's vast resources will be the regional countries, assuming that continued melting of the ice makes way for its full exploitation. Nevertheless, certain non-regional states (China, India, Japan, and South Korea), as well as the European Union (EU), also have stakes in those resources, not as the owners but mainly as beneficiaries of the resources to be available for global use through large-scale regional exports. This is especially true of the region's oil and gas, as their availability could address the energy security concerns of states having large and growing energy requirements that must be met in large part by imports. The rise of the Arctic as a major oil and gas supplier will certainly provide them with an additional large supplier and help them further diversify their sources of supplies and also secure reliable resources for their increasing energy imports. The possibility of two sea routes via the Arctic, which are shorter than the Southern Sea Route currently in use is yet another reason for the mentioned states' stake in the Arctic as they are all large trading countries. Section III elaborates on the importance of the resources and routes for certain non-regional states (India and Japan) and the EU. Membership of Denmark as an Arctic state through Greenland, and of Sweden and Finland as Greater Arctic states in the EU provide additional interest in the region for Brussels.

Neil Gadihoke sheds light on India's interest in the Arctic, specifically of course, to its expanding energy imports as its domestic energy resources are unable to match its increasing energy demand. Importing oil and gas from the Arctic is, therefore, a promising prospect for the Indians who want to ensure the availability of supplies even under the worst case scenario requiring further diversification of suppliers. India is also closely watching the situation in the Arctic for its potential sea routes, but its interest in the routes is not linked to use of them for trade. Indeed, it is more concerned with loss of income due to the diversion of some of sea traffic from the southern route to the northern ones (NSR and NWP). In addition, rising sea levels as a result of the Arctic's melting ice is certainly a major reason for India's interest in the region. As a country with a long coastline along the Indian Ocean, the melting, if it continues, will have dire consequences for the adjacent areas, including those of its neighbour, Bangladesh, from which millions of people could flow into India in search of safety and a livelihood, a scenario with both negative economic and security implications for India. Consequently, Gadihoke makes a case as to why India is an interested party in the Arctic, despite its status as a non-regional state, and its current engagement in that region being confined to scientific expeditions.

Like India, Japan is an interested non-regional country in the Arctic. According to Hiroshi Ohta, Japan's main involvement, up to this point of time, in the Arctic has been confined to various environmental and scientific exploratory expeditions. These activitities signify Japan's view of the melting ice as a major threat to global environmental health. It is also a source of concern for the Japanese because, as an island country surrounded by sea waters, Japan will suffer extensively if sea levels continue to rise, a scenario that they want to prevent from realization. Hence, environmental security is part of Japan's understanding of security known as comprehensive security. Ohta explains why for Japan the traditional concept of security, mainly based on military security, is not applicable even though military security is surely part of its comprehensive security. Yet, apart from environmental concerns, Japan is certainly interested in accessing the Arctic's oil and gas through imports. Having almost no domestic oil, gas and coal reserves, the country is heavily dependent on large imports of these fuels for which supply security is a major source of concern. Accordingly, any major interruption in its imports of fuel would have serious economic implications for Japan as a major global exporter of goods, and also disturb the Japanese way of life. Unsurprisingly, the availability of large-scale oil and gas from the Arctic could allay Japan's energy security concerns by providing it with an additional major supplying region. Thanks to its geographical location, Japan is a beneficiary of the sea routes via the Arctic through which it could conduct at least part of its international trade at a significantly lower cost, another reason for its interest in the Arctic.

The EU is of course interested in the Arctic for various reasons. According to Danila Bochkarev, one EU member — Denmark — is an Arctic state through Greenland, but Greenland's semi-autonomous status and its decision to stay out of the EU framework have weakened its access to the region. Its other two members, Finland and Sweden, are part of the Greater Arctic region, with very limited claims to the Arctic due to their geographical location. However, the EU's interests in the Arctic are only partly tied to some of its membership's geographical location. In fact, the proximity of its membership to the region makes it interested in it though the EU is still kept outside of the regional governing bodies with the effect of denying it a strong say in the regional affairs. As a regional grouping with limited oil, gas and coal resources within its boundaries, the EU is heavily dependent on imported fuels especially oil and gas. Its dependency on Russia for these has been steadily increasing, a trend of concern for Brussels because of its potential political and security implications. Thus, decreasing its reliance on Russia, and other current suppliers, and instead importing large amounts of oil and gas from the Arctic is a very interesting prospect for the EU. Similarly, the vast mineral reserves of the Arctic are important to the EU because its membership's own mineral resources have been depleting, necessitating large importation from outside the EU. The prospect of conducting trade via the NSR and the NWP is surely another major reason for the EU's interest in the Arctic. Heavily dependent on exports, the EU's key members could benefit from such routes, which could help them improve their competiveness by decreasing the cost of their exports at a time when they are facing the challenge of many new exporting nations, particularly in Asia. As the EU's fish stocks are also depleting, the Arctic is similarly attractive in that regard. Thus, Bochkarev stresses the importance of the Arctic's fossil energy reserves, mineral deposits, sea routes and fish stocks as four reasons for the EU's interest in the Arctic, though these are not the only reasons. The region's environmental degradation and the rising sea levels are of major concern due to their dire consequences, not only for nearby Europe, but for the entire planet. For all these reasons, the EU is and will remain interested in the region, but it will not be an agenda-setter at least in the foreseeable future, due to its membership's limited access to the region, which however, could change if Greenland opts to join the EU.

***

The mentioned arguments presented in detail in the previous chapters clarify the implications of the Arctic's melting ice for all countries, because of its direct and indirect consequences for the global environment, and the global energy and mineral markets as well as global trade, and therefore, the global economy. Of course, those states with direct access to the region, due to their geographical location, will be the main beneficiaries of the economic gains, including the vast oil and gas resources. They will also be the first to suffer from the negative outcomes, such as rising sea levels. The root cause of either situation is of global concern given the importance of the Arctic for the survival of the Earth in its current form. Surely, the melting of the Arctic ice, if it is not stopped, will be a disaster for our planet. It is caused mainly by global warming, whose main source is now widely believed to be the unsustainable pattern of our energy consumption based heavily on CO_2-emitting fossil fuels, even though some scholars attribute it to other natural factors. To deal with this issue, it is necessary to sharply decrease our consumption of fossil energy. Unfortunately, despite a relatively significant expansion of the non-fossil energy sector, the global energy mix is still dominated by fossil fuels (86.9% in 2010) and these fuels are projected to continue to dominate (73% in 2035) if the current pattern of energy consumption continues.[2] At the time of writing, there was no strong evidence to suggest a change in this reality in the foreseeable future.

Consequently, the melting of the Arctic ice will continue. This environmentally negative phenomenon will pose major challenges, including the worsening of global warming by denying the Earth its major cooling system. The rising sea levels will affect all countries with coastlines along open seas. The melting of the ice will also offer opportunities by making large-scale oil and gas extraction and mining possible, and also provide sea routes via the Arctic that will be available for a much longer period of time than it is now. As a result, in the short-and medium-term, thus before the full extent of the negative repercussions seriously challenge the continuity of our life in its current form, the melting of the ice will offer the mentioned opportunities. Based on this assumption, certain trends are identifiable reflecting the interests of the countries with direct natural access to the Arctic and those of the major non-regional economies with a capability engaged in the region.

One is a growing engagement of the Arctic littoral states. In the foreseeable future, most efforts will be focused on securing access to and developing the region's resources, mainly its oil and gas, for which there will be no shortage of customers in the region and beyond, thanks to the growing global demand for energy.

Another is the indirect engagement of the major Asian economies (China, India, Japan, and South Korea) and the EU in the development of these resources. Developing the region's oil and gas reserves is very capital-intensive, especially those located offshore. Although there are differences in terms of availability of investment between and among the Arctic states, no one state has sufficient financial resources to undertake all the required development projects on its own. The mentioned Asian economies have large funds to position them as major foreign investors in the Arctic oil and gas development projects, but given the current and expected continuation of the serious economic/financial situation of most of the EU's members, the EU's involvement will likely be more tied to supplying technology.

Russia will be the main player in developing the Arctic resources within the currently recognized boundaries, although all the other states in the region will take measures to expand their roles. At least based on the publically available information, Russia has the largest share of the Arctic's proven oil and gas reserves and has already taken major steps to develop them. Moreover, it is far better equipped than the other regional and non-regional states to access the region thanks to its having the world's largest icebreaker fleet, which is currently expanding.

A significant trend in the Arctic is the participation of the non-littoral large economies. Although China has acquired an icebreaker in order to boost its position compared to other non-regional states in terms of accessing the Arctic, it is not very likely that China will seek to extensively expand its presence in the region to develop the offshore oil and gas resources in the areas falling outside the littoral countries' exclusive economic zones. Nevertheless, it will certainly expand its current presence by holding more scientific expeditions. To avoid major conflicts, especially with the militarily-strong Arctic countries (Russia and USA), when its economy needs a few more decades of sustained economic growth to address the country's current shortcomings, China's efforts towards direct access to the regional oil and gas resources in undisputed areas will be limited. Beijing will certainly seek to secure a large share of the regional energy exports through large investments in their respective fields. As well, other non-littoral states' engagement in the regional mineral and energy resources will likely be mainly in the form of investments to secure access to the resources, particularly oil and gas, through imports. Nonetheless, they will all make an effort to expand their presence through more expeditions.

A gradual militarization of the region is a distinct trend. All of the countries in the region have already taken steps to beef up their regional military presence, but most of them simply lack the resources to emerge as major military players. Russia and the United States are two exceptions due to their long military presence in the Arctic and their military capabilities. The two powers will certainly strengthen their regional military presence, but a major arms race between the two, at least in the foreseeable future, is unlikely due to various factors, including their both suffering from major economic problems, which are limiting their resources for a large-scale military build-up. As for the non-regional states, their military presence will be mainly symbolic because they lack the required assets and have no strong reason for a military build-up in the Arctic.

Finally, related to the former, the economic stakes are high in the Arctic, but the exploitation of the region's resources will be difficult and costly due to the harsh environment and the technological complexity of accessing the offshore resources, in particular. Moreover there is opposition, arising from environmental concerns, in all the Arctic countries to extensive development of the Arctic resources. Indeed, this factor is delaying such development by most of the littoral states. The escalation of disputes over the undivided Arctic areas into major political conflicts, let alone military ones, will be unlikely but such disputes will probably surface over time. At least for a predictably long period of time, the littoral states will focus on the resources in their undisputed areas.

In conclusion, the melting of the ice in the Arctic is certainly a major threat to the environmental health of Planet Earth, serious enough to pose major challenges to the survival of all forms of life.

As it stands today, the only way to reverse the process is to address global warming. This requires a sharp reduction in GHG emissions. Despite the efforts to reduce such emissions over the last two decades, the targeted levels are too low to reverse the process as they are set to maintain global warming within the "acceptable" level regardless of the apparent negative consequences (e.g., forest fires, rapid depletion of fresh water resources and decreasing agricultural yields). One hopes that the sheer scale of the catastrophic natural events resulting from a major or a complete melting of the Arctic's ice will convince all countries to take measures to reverse global warming. The comparatively short-term economic gains of the disappearing ice will be too small to justify such devastating results.

BIBLIOGRAPHY

BP (2011). Primary energy consumption. *BP Statistical Review of World Energy 2011*, London: BP.

IEA (2010). New policies scenario. *World Energy Outlook 2010*. Paris: IEA.

ENDNOTES

[1] BP (2011), Primary energy consumption, *BP Statistical Review of World Energy 2011* (London: BP), p. 40.

[2] IEA (2010), New policies scenario, *World Energy Outlook 2010* (Paris: IEA), p. 620. Author's calculated percentage of fossil energy of the global energy mix in 2010 based on the data available in BP (2011), p. 41.

INDEX